SpringerBriefs in Agriculture

For further volumes:
http://www.springer.com/series/10183

Robert L. Zimdahl

Weed Science: A Plea for Thought—*Revisited*

Prof. Robert L. Zimdahl
Bioagricultural Sciences and Pest Management
Colorado State University
Campus Delivery 1177
Fort Collins, CO 80523-1177
USA
e-mail: r.zimdahl@colostate.edu

ISSN 2211-808X   e-ISSN 2211-8098
ISBN 978-94-007-2087-9   e-ISBN 978-94-007-2088-6
DOI 10.1007/978-94-007-2088-6
Springer Dordrecht Heidelberg London New York

Cover design: eStudio Calamar, Berlin/Figueres

Printed on acid-free paper

Springer is part of Springer Science+Business Media (www.springer.com)

Abstract

The original essay was written in 1991 and published by the US Department of Agriculture in 1993. After twenty years it is appropriate to ask if weed science has overcome the paralysis of the pesticide paradigm and if the discipline's research emphases have changed. Weed scientists are confident of increasing production through intelligent use of agricultural technology, including herbicides. But, we must ask if the moral obligation to feed people is sufficient justification for the benefits and harms achieved. A continuing, rigorous examination of the science's goals that leads to appropriate changes is advocated. People agree that all goals and the means to achieve them should be good. Inevitable value questions arise because people do not agree on what is good, true or on what ought to be done. There is little public consensus about the necessity and value of widespread pesticide use to increase food production and improve public health. Weed scientists have a research consensus, and thus a paradigm, which should be explored. The paradigm has two propositions: 1. there are weeds that must be controlled and 2. herbicides are the primary, most efficient control technology. Since 1800 the indisputable evidence shows that agriculture has contributed significantly to the fact that the majority of the earth's population is better fed, better sheltered, protected from disease, richer and lives longer. This perception of success affects how agriculture is practiced in developed and developing nations. The conventional wisdom is that herbicides are necessary tools of modern technology avocated by nearly all parts of the agricultural enterprise. Agriculture's practitoners should engage in regular discussion of the necessity and risks of all pesticides for continued agricultural progress. These will not and should not focus only on the scientific evidence. They will include and must address value-laden arguments. Separating issue of fact from issues of value is fundamental to debate about weed science's future.

Keywords Collaboration · Controlling knowledge · Definitions · Ecologists · Entomology · Facts · Future · Goals · Herbicides · History · Historical reflection · Morality · Plant pathology · Paradigm · Pests · Pesticides · Plant protection disciplines · Plea · Prometheus myth · Receiving knowledge · Research · Risk · Shifts · Values · Weed control · Weed science · Weeds

Preface

Hope resides in the future, while perspective and wisdom are almost always found by looking to the past.

Mortenson G. 2009, p. 21

The past is a foreign country. They do things differently there.

Hartley L. P. 1967, p. 3

Weed science—A plea for thought was written in 1990 and published by the Cooperative State Research Service[1] (CSRS) of the US Department of Agriculture in 1991 as a symposium preprint. The symposium was held in Washington D.C. on 15 April 1993. It was convened by J. P. Jordan, Administrator of CSRS/USDA. I was the first speaker. Other speakers and their affiliation in 1993 were: John Abernathy, Texas A&M University; David Bridges, University of Georgia; Harold Coble, North Carolina State University; Jodie Holt, University of California—Riverside; and Donald Wyse, University of Minnesota.[2] F. Dan Hess, Sandoz Agro, Inc., submitted a manuscript.

Dr. John A. Naegele, in the preface for the essay published in 1991, noted that the essay asked if the weed science research planning community could "overcome the paralysis of the pesticide paradigm and conceive a weed science research program that addresses both society's perception of safety and the scientific community's perception of risk?" The essay was to serve as a "cognitive launching pad" for a CSRS sponsored weed science research planning symposium to be held in 1993.

After 20 years it is appropriate to revisit my thoughts from 1990–1991. It is reasonable to ask if the weed science community has "overcome the paralysis of the pesticide paradigm" and if weed scientists and the discipline's research emphases have changed their focus and goals. Have weed scientists thought about

---

[1] CSRS is now The National Institute of Food and Agriculture (NIFA).

[2] The papers were published in Weed Technology 8:388–412. 1994

the direction and goals of their science and whether they are acceptable or need to be modified? Have they developed reasons to explain why modification is or is not necessary? 20 years ago in the enormously successful agricultural system in the developed world the dominant weed control techniques clearly had great (sometimes complete) reliance on herbicides. Weed and other agricultural scientists were confident of the truth of their basic faith in the possibility of perpetually increasing production through intelligent use of ever more efficient agricultural technology, including herbicides. The justification was the moral obligation to feed people. Is this belief still prevalent? Is it justified by the evidence and rational argument? The original essay was not intended to demean, diminish or be only critical of the great accomplishments of weed science. It was a plea for thought, as is this re-visitation. Knowing about the mistakes and successes of the past is vital, not to return to the past, but to learn from it (Cox 2009, p. 57).

To retain the intent and integrity of the original essay nearly all of it has been retained including the chapter titles and literature cited. Chapters 2 and 3, with minimal changes appear as they did in the 1991 essay. The essential message of Chaps. 1 and 4 has been retained, but both have several additions. Chapter 5—The Future is a new, brief, conclusion. Editing has been primarily to correct mistakes, to acknowledge the passage of 20 years, and include knowledge gained in those years.

Sen. Daniel Patrick Moynihan is purported to have said: "We each may be entitled to our own set of opinions, but we are not entitled to our own set of facts." In the original and this revision I have tried to get the facts right. If I have not succeeded, please inform me of my errors. The opinions are, of course, my own. The purpose of this essay is identical to the purpose of the original. It is to plead for thought about who we are, where we have come from, where we are going and where we ought to go. If the essay accomplishes or fails to accomplish these goals, I look forward to hearing and will appreciate knowing your view.

Literature Cited

Cox H (2009) The future of faith. Harper Collins, New York, pp 245

Hartley LP (1967) The go-between. Stein and Day, New York, pp 311

Mortenson G (2009) Stones into schools—promoting peace with books not bombs in Afghanistan and Pakistan. Penguin books, Inc. New York, pp 420

Naegele J (1993) See Preface, Zimdahl RL (1993) Weed science—A plea for thought.

Acknowledgments

The 1991 version of this essay was written without the aid of reviewers. This revision has not been hampered by the same error.

Many of the thoughts and arguments in the essay were created and nurtured over many years in classes I have taught and others I have attended during seminars, professional meetings, and in numerous conversations with colleagues. All of these took place over several years and with far too many people to list them all. I am indebted to all who have given me the privilege of sharing their thoughts, even when we knew we did not agree.

Dr. Thomas O. Holtzer, Professor and Head, Department of Bioagricultural Sciences and Pest Management, Colorado State University has supported and advocated my work by providing office space and departmental administrative support. His careful reviews of portions of the manuscript have improved clarity and presentation of ideas. Dr. K. George Beck, Professor, Department of Bioagricultural Sciences and Pest Management, Colorado State University critically reviewed portions of the manuscript. Members of the Publication Coordination Committee of the Weed Science Society of America provided a brief, useful comment.

I extend special appreciation to Dr. James W. Boyd, Professor Emeritus of Philosophy, Colorado State University who remains a valued friend and philosophical mentor. I express my gratitude to Ms. Maggie Hirko and Ms. Janet Dill who have made my task more pleasant and easier by their courtesy, regular assistance, and tolerance of my, perhaps too frequent, requests for assistance.

The inside cover picture is used with permission of the artist, Jim Foster, Waverly, CO, fostart.jimart@gmail.com

My wife, Pamela J. Zimdahl, encouraged my writing and offered comments and criticism when she thought it was appropriate (it usually was).

March 2011   R. L. Zimdahl

Contents

Chapter 1
The Need for Historical Perspective

Prometheus, the Titan, employed by Zeus to make men out of mud and water, stole fire from the heavens and gave its power to man. With the gift of fire, Prometheus fulfilled his destiny to be creative and a courageously original life-giver. With fire, man had the power to become toolmaker, explorer, and food grower. However, Zeus did not appreciate disobedience by a lesser god and punished Prometheus by chaining him to Mount Caucasus where an eagle came each day to eat out his liver, which Zeus renewed each night.

Although Prometheus was punished, the gods schemed further to control man's new power. They created Pandora whose name means "all gifted" because each of the gods gave her something. She was endowed with Aphrodite's beauty, Hermes' gift of persuasion, and Apollo's music to entice the heart of man away from full use of fire's power. They also gave Pandora burning curiosity, as well as a box, a gift for the man she would marry. With the box came the warning that it must never be opened. Prometheus (whose name means forethought) mistrusted Zeus and his gifts, but his brother Epimetheus (afterthought) married the beautiful Pandora, and accepted the box. One of them, Pandora or Epimetheus, opened the box, and once opened, it could not be closed; thus, the story tells us, all that is evil escaped into the world to torment mankind forever, only one thing, hope, remained in the box. It is interesting that Prometheus—forward looking, life-giving, creative, courageous, and Pandora—beautiful, enticing and persuasive, but whose fatal curiosity loosed a thousand plagues, are part of the same myth!

Myths are traditional stories many of which originated in pre-literate societies. Thus, in the strict sense they are not true. They are best regarded as pointing toward truth. Myths appeal to the consciousness of people by expressing cultural ideas that give expression to deep, commonly held emotions. I choose to begin with the Promethean story because it helps me think about who we are and where we have come from. For me it stimulates forward-looking, creative and courageous thought. For most of modern history, the western world has enjoyed its Promethean power—the power of science. It has permitted us to evolve from makers of simple tools to

R. L. Zimdahl, *Weed Science: A Plea for Thought*—Revisited,
SpringerBriefs in Agriculture, DOI: 10.1007/978-94-007-2088-6_1,
© Robert L. Zimdahl 2012

developers of sophisticated instruments and machines and from explorers to conquerors. The power of science transformed the developed world's agriculture from subsistence to abundance and even surplus. We quickly learned how to use and benefit from the Promethean gift of fire and thus, power. Pandora was forgotten except as part of an interesting myth. But, as Kirschenmann (2010) points out, although the gift of fire allowed us to achieve domination, we have not achieved dominion nor has our immense power given us control over the natural world. For all its wonders and undeniable benefits, science and its associated technology have a disquieting aura of fallibility. The results of Pandora's curiosity may be a more important part of our inheritance than Prometheus' gift.

While the Prometheus story helps us think about where we have come from, the following story by Loren Eiseley, a respected anthropologist and distinguished writer, has helped me think about where we are going. Eiseley told of his experience when he boarded a train in the confusion of a New York weekend. His story demands thought and, in my view, could assist us as we, in Eiseley's words, "confront reality and decide our pathway".

> My profession demands that I be alert to the signs and portents in both the natural and human worlds - events or sayings that others might regard as trivial but to which the gods may have entrusted momentary meaning, pertinence, or power, such words may be uttered by those unconscious of their significance, casually, as in a bit of overheard conversation between two men idling on a street, or in a bar at midnight. They may also be spoken upon journeys, for it is then that man in the role of the stranger must constantly confront reality and decide his pathway. It was on such an occasion not long ago that I overheard a statement from a ragged derelict which would've been out of place in any age except, perhaps, that of the Roman twilight or our own time. A remark of this kind is one that a knowledgeable Greek would have examined for a gods hidden meaning and because of which a military commander, upon overhearing the words, might have postponed the crucial battle or recast his augueries. I had come into the smoking compartment of a train at midnight, out of the tumult of a New York weekend. As I settled into a corner I noticed a man with a paper sack a few seats beyond me. He was meager of flesh and his cheeks had already taken on the molding of the skull beneath them. His threadbare clothing suggested that his remaining possessions were contained in the sack poised on his knees. His eyes were closed, his head flung back. He drowsed either from exhaustion or liquor, or both. In that city at midnight there were many like him. By degrees the train filled and took its way into the dark. After a time the conductor shouldered his way in, demanding tickets. It is thus one hears from the gods. "Tickets! Bawled the conductor." I suppose everyone in the car was watching for the usual thing to occur. What happened was much more terrible. Slowly the man opened his eyes, a dead man's eyes. Slowly a stick-like arm reached down and fumbled in his pocket, producing a roll of bills. "Give me," he said then, and his voice held the croak of a Raven in a churchyard, "Give me a ticket to wherever it is."

In a single sentence that cadaverous individual had epitomized modern time as opposed to Christian time and in the same breath had pronounced the destination of the modern world.

Eiseley (1971) pp. 62–63

Give me a ticket to wherever it is! Weed and nearly all agricultural scientists have known where they are going. We have been sure of our destination and, if necessary, justifying it was easy. The reasons for the journey were clear, albeit unexamined. Only rarely have we paused to consider if where we are going is where we ought to go and if our reasons for continuing toward the same destination are the right reasons.

Science, including weed science, has given to no one in particular, powers that no one knows enough about. Our Promethean power has permitted great achievements, but it has not made it easier for us to know wherever it is we are destined to go, to do. Since the 19th century, communication speed probably increased more than 1,000 times. We have moved from sea-mail and the pony express to nearly instantaneous, worldwide, satellite-aided communication. Who can get along now without e-mail and the internet?

When Levi Zendt left his Amish heritage and traveled west to Colorado in James Michener's novel "Centennial" his wagon, pulled by horses, could make about 10 miles per day, on level terrain in good weather. Now we think nothing of flying across the country in half a day or driving several hundred miles in a day. Spaceships which travel 18,000 miles per hour are no longer front-page news. In World War II, a blockbuster bomb containing 20 tons of TNT could destroy a whole city block. The Hiroshima atomic bomb had the explosive equivalent of 20,000 tons of TNT and now a single thermonuclear hydrogen bomb has the explosive equivalent of 25 million tons of TNT. That is more explosive power, in a single bomb, than in all the bombs dropped, by both sides, in World War II. Our ability to destroy has increased one million times in the last 40 years.

Almost every citizen of developed countries has experienced the benefits of modern antibiotics, which started with Alexander Fleming's discovery of penicillin in 1929. Modern birth control technology has made life easier for many women and families. In the 1940s my parents worried about infantile paralysis (polio), but I never had to worry about its effects on my children, thanks to Dr. Salk and Dr. Sabin. Smallpox has been eradicated from the world. In the last 50 + years a series of discoveries in molecular biology have revealed the molecular structure of proteins, nucleic acids, DNA and our genes. Scientists now modify genetic structure and perhaps will create new genomes. Your great grandchildren may, one day, order designer genes from a catalog. They surely will not worry about the diseases we worried about for our children. Continued research will reveal the cause(s) and may very well provide a cure for devastating diseases e.g., cancer, amytrophic lateral sclerosis (ALS), AIDS/HIV, and malaria. In the US parents are no longer concerned about measles, chickenpox, diphtheria, and mumps. Immunization has taken care of them. This is not true in the world's developing countries where many of these diseases are endemic.

Modern science is the engine that drives the West's great agricultural productivity. It is the key to power and an important part of the solution to problems faced by agriculture, many of which were caused, or exacerbated, by technological achievements developed by scientists. Science cannot solve all problems, but many cannot be untangled without it. American agriculture has achieved its undeniable productive success by substituting scientific knowledge for resources (Ruttan 1986). The knowledge has been demonstrated in mechanics, biology, and chemistry and in farmers' managerial skills (Ruttan). Since 1900 world population has increased 400%; crop land 30%; average crop yield 400%; and total crop harvest 600% (Ridley 2010, p. 143). World agriculture now provides food for more than 7 billion people, more than ever before. We often like to think that world agriculture means farmers actually grow food. That is false. World agriculture includes farmers (now <1% of the US population) that produce crops and animals for the food system. The food system also includes: purchasers, wholesalers, retailers, processors, packagers, transporters, manufacturers, and the range of manufacturers and suppliers that provide things modern agriculture requires: pesticides, fertilizer, seed, fuel, medicine, feed (Berry 1981, p. 115). It is that system that permits the claim that citizens of developed countries are better fed, earn more money, are better sheltered, much better entertained, protected against what used to be common diseases, and much more likely to live beyond three score and ten compared to the average person at the beginning of the 19th century (Ridley 2010, p. 12). By the 1960s, improvements in all agricultural technologies had changed the world food situation from shortage to surplus. Conversely, in this land of plenty 25% of the population is overweight or obese and 4% of US citizens do not have a secure food supply. One billion of the world's people still lack adequate food and safe water, but these problems are commonly regarded as being caused by a combination of social, cultural and governmental effects. That is they are not regarded as agricultural production problems. No one denies the reality of the problems, but they are neither caused by nor will they be solved by agricultural scientists. It is not their task. It is not what they were trained to do or are capable of doing. However, the highly developed monocultural agricultural food production system developed by agricultural scientists and used extensively in the world's richer nations has caused numerous problems (e.g., loss of biodiversity and crop genetic diversity, diminished soil health, loss of ecosystem services). When these problems are combined with the looming end of cheap energy, declining freshwater resources, global climate change leading to unstable climates, and an expanding human population intent on increasing consumption of food and resources, it is difficult to deny that we face challenging problems and that agriculture has a role to play in solving them.

Development and use of synthetic fertilizers and pesticides has been at the heart of advances in agricultural productivity, which could not have occurred without great improvements in plant productivity contributed by selective plant breeding. The green revolution was, in large measure, a contribution of modern plant breeding and its combination with pesticides, fertilizer, and irrigation. Pesticides of all kinds have permitted modification of crop environments by reducing or eliminating insect, disease and weed populations. Modern agricultural technologies have permitted humans to shape the environment to meet their needs and wants. In spite of all its important achievements, agriculture remains an unappreciated and often neglected human activity. It is the world's largest and most important environmental interaction, albeit, one with important negative effects. Agricultural scientists and food producers accept credit for their role in feeding people, a morally good thing. Scientists are quite aware of the problems, mentioned above, and those of the dominant monocultural production system. However, these problems while generally acknowledged to be important, are not important enough to supersede the quest for continued production. They are a result, albeit undesirable, of the perceived necessity of using modern agricultural technology to produce enough food for the world. Weed science illustrates the modern dilemma of dependence on science and the untoward effects of that science. It is the role of herbicides in agriculture and weed science that will be dealt with herein. The essay will explore some aspects of the history of weed control and weed science (these are not synonyms), and some ensuing, perhaps inevitable, problems from their widespread use while reflecting on the future of weed science and pesticide use in general.

Historical Reflection

Technologists do not like historical reflection, especially if it is critical of what is viewed as progress toward human development, a better life, growing production with greater efficiency, increased profit, etc. Technologists seek results and progress toward a better life for all. In the technological realm, it is the goal and not necessarily the route to achieving it that dominates thought (Knusli 1970). The computer at which I am typing is a good example. My first one arrived in 1987, before it was 2 years old it had become an antique. It still worked and did all I wanted or had the ability to do, but new ones did so much more. Our lives are increasingly ordered by these advances. Not too long ago, one could actually do research, conduct business, and carry on the chores of daily life without worrying whether the computers were up or down. People did things without the aid of modern technological marvels. Many people now begin the day by checking their e-mail, which was unknown prior to 1971.[1] Many believe they could not do things well without their cell phone, which was invented in 1973.[2] It is likely you and nearly all of your friends have at least one.

Few technologists ask if their achievements are innately desirable or if they will have desirable social consequences beyond their intended function. It seems that technological progress is the end, the achievement, and that it is accepted almost without question. The technological challenge is to grow, to get bigger, and importantly, to aid social and individual progress toward a better life for all. Similarly, weed scientists have not been pressed or even encouraged to explore the many implications of the technology that has aided their undeniable achievements, their science's progress. When increased production and positive economic effects result, the evidence is slim that weed scientists carefully analyze other possible consequences (i.e. cultural, social, ecological) of the achievement. It is true that herbicides with undesirable human health or environmental effects are not going to reach the market as some did a few decades ago. An aspect of science, including weed science,

---

[1] The text of the first electronic missive consisted of "something like QWERTYUIOP." It was sent by computer engineer Ray Tomlinson in 1971. It was simply a test message to himself, sent from one computer to another sitting right beside it in Cambridge, Massachusetts. See HTTP://ask.yahoo.com/20010824.HTML—Accessed October 27, 2010.

[2] Martin Cooper invented the first portable hand phone (the predecessor of today's cell phone) in 1973, Today there are 5 billion in the world. 285.6 million (91%) Americans had cell phones in 2010.

is the unexamined intellectual conflict of interest that pressures scientists to pursue the research that is most likely to lead to governmental or industry funding,[3] rather than that which their culture, society, or ecological environment tells them should be done. Weed scientists similar to other scientists are certain of their incorruptibility and the independence of their thinking and research from the source of the funds that enable the research. Scientists acknowledge that grants or gifts from the agro-chemical industry affect research topics, but deny that they affect interpretation of results. There is no question that, independent of the source of research funding, the "scientific enterprise is probably the most fantastic achievement in human history" (Freedman 2010). Science is "more than an instrument of man's increasing power and progress. It is also an instrument, for the malleable adaptation of man to his environment and the adjustment of his environment to man" (Glass 1965). But its achievements do not mean that scientists have a right to overstate what they've accomplished or to claim they are never influenced by the source of the money that funds their research program (Freedman, p 86).

Weed scientists, and other pest scientists, point with pride to their achievements. Disease-and insect-free (or nearly free) crops can be produced in a weed-free environment. Crops yield more and are purported to be more profitable for growers and are claimed to be more healthy, or at least not unhealthy, for consumers. Achieving a weed free cropping system has been aided by the development of genetically engineered crops. The best example is development of Roundup Ready™ crop seed by Monsanto. The agricultural plants created by genetic modification, are resistant to the herbicide Roundup™ which, when first introduced, killed nearly all plants; crops and weeds.

Those in pest control have focused on solving the technical/production problems created by pests which reduce crop yield and lower quality. The focus has been to reduce pest incidence. It has not been to reflect on means, but to accomplish the goal of improved pest management and as new technology becomes available, to use it for further improvement. Knusli (1970) contrasts scientists in pest control with a bridge-building engineer who, one day, will stand by the construction "tested, safe, ready to use, with no question left open". Although the assumption of "no further questions" is often wrong, the engineer moves on to the next, presumably better project which will benefit from cumulative experience. The last project, it is assumed, was complete and final. Knusli goes on to point out that the community of natural scientists cannot dream of reaching a point where a project is complete and final in and of itself. Weed scientists, as part of the larger community of natural scientists, also cannot dream of reaching a point where their science is complete and final: no further improvement is possible. Natural scientists are building an expanding, deepening picture of nature that will never be complete. They build an evolving view of the

---

[3] It is reasonable to argue that government funding priorities in a democratic society represent the will of the people. It is also reasonable to claim that industry funding of research follows market forces which, in a free market system, reflect societal priorities. One could also claim that priorities are heavily influenced by those with money. Discussing these competing claims is beyond the scope of this essay.

natural world. It is an image from a particular point of view and place in time, not the final answer. The view is a reflection of the idea that "We see the world, not as it is, but as we are or, as we are conditioned to see it" (Covey 1989[4]).

Most biologists accede to the view that their research is contributing to an expanding picture of nature that will never be complete. However, in some sectors of biology, I think especially in weed control, scientists may not operate from this broader biological perspective. We know weed control is evolving, but its evolution has been constrained because 20 years ago the science focused almost exclusively on a single solution to the problem. The desirable goal of weed control was too frequently hitched to the technological achievement of herbicides. Decreasing emphasis on herbicides and increasing emphasis on the ecological and biological aspects of weeds is indicative of desirable change.

What is a Weed?

Weeds are ubiquitous, common, and bothersome plants that have been described in terms of their habitat, their behavior, their undesirability, and their virtue or lack thereof. However, not all agree on what a weed is or which plants are weeds. For example:

"A plant out of place or growing where it is not desired." Blatchley (1912)

"Any plant whose virtues have not been discovered." Emerson (1876)

"Any plant other than the crop." Brenchley (1920)

"A plant not wanted and therefore to be destroyed." Bailey and Bailey (1941)

"Those plants with harmful or objectionable habits or characteristics which grow where they are not wanted, usually in places where it is desired something else should grow." Muenscher (1960)

"A very unsightly plant with wild growth, often found in land that has been cultivated." Thomas (1956)

"Weeds are pioneers of secondary succession, of which the arable field is a special case." Bunting (1960)

"A plant is a weed if, in any specified geographical area, its populations grow entirely or predominantly in situations disturbed by man." Baker (1965)

"A herbaceous plant not valued for use or beauty, growing wild and rank, and regarded as cumbering the ground or hindering the growth of superior vegetation." Little et al. (1973)

"A weed is a plant that originated in a natural environment and, in response to imposed or natural environments, evolved, and continues to do so, as an interfering associate with our crops and activities." Aldrich (1984)

---

[4] This aphorism is frequently and incorrectly attributed to the Talmud. Its source is quotes from S. Covey (HTTP://www.goodreads.com/author/quotes/1538.Stephen_R_Covey, Accessed January 2011). Similar words are found on pages 17 and 32 of Covey 1989.

These definitions range from the poetic descriptions of Emerson to the didactic terms of Baker, Bunting, and Aldrich and the agronomic, control-oriented of Blatchley, Brenchley, Muenscher, and Bailey. The Weed Science Society of America (Anonymous 1983) defines a weed as "any plant that is objectionable or interferes with the activities or welfare of man." These definitions are accepted by practitioners of weed control who rarely ask how close they are to truth.

The Role of Definitions

It is important to note that man is the focus of most definitions of weeds. There is no botanical or taxonomic category of weeds, although there is a clear definition of invasive plants.[5] We humans say that a plant, at a certain time and place, regardless of its origin, is objectionable or interferes with our activities. It often does not matter that at another time and place it could be desirable or at least not objectionable. That is changing with weed science's increasing emphasis on invasive species. The definition is not a personal opinion. For example, Kentucky bluegrass (*Poa pratensis* L.) is desirable in lawns and parks, but not in Rocky Mountain National Park. Downy brome grass (*Bromus tectorum* L.) (often called cheat grass), an invasive common weed on western rangeland is usually undesirable, but is considered reasonable as a winter and early spring feed for rangeland livestock and large ungulates.

The definitions of weeds create a view of nature that is common, if not ubiquitous among weed scientists (and perhaps among most practitioners of pest control). The definitions imply—once a weed, always a weed. There is no presumption of innocence about weeds, whereas, dependent on location, there often is about invasive species. Only Aldrich (1984) offers a definition of weeds that provides, as he puts it, "both an origin and continuing change perspective". He says that "recognizing that weeds are part of a dynamic, not static, ecosystem helps us to expand our thinking on how best to prevent losses from them". His definition moves away from those that regard weeds as enemies to be controlled. It is an ecological definition that regards weeds as plants with particular, perhaps unique, characteristics and adaptations that enable them to survive and prosper in disturbed environments.

Although all do not agree on what a weed is, most people know they are not desirable. Many even recognize the human role in creating the negative image weeds have. Often it is a richly deserved image because weeds are detrimental and must be controlled, but is this always true? Is their lack of virtue more a function of the image created by our definitions than an absolute fact of nature? For

---

[5] An invasive species is non-native (or alien) to the ecosystem under consideration and whose introduction causes or is likely to cause economic or environmental harm or harm to human health. (Presidential Executive Order 13112 Feb 3, 1999).

example, when the environment is disturbed by fire, flood, war, or construction it is common that the first plants to appear are weeds. In a metaphorical sense, they act as an ecological Red Cross. Weeds appear to heal the wounds, help the injured re-establish, protect the soil and begin to restore the environment. Our attitude and our definitions shape our lives and relationships with other things. Another frequently encountered definition of weeds says they often occur in waste places. Weed scientists have heard and used this definition for years without questioning the term waste place. We assume it is a place where crops are not grown and plants grow wild and cumber the ground and where it would be good if something else would grow—perhaps something pretty, or at least not ugly. The fact that we accept that weeds occur in waste places (abandoned areas) creates an attitude toward them and all with which they are associated.

Definitions are powerful creators of images and meaning and we cannot escape them through pleas of objectivity or freedom from values. Capra (1982) reminds us that "patterns scientists observe in nature are intimately connected with the patterns in their minds, with their concepts, thoughts, and values". We see things as we are. The research weed scientists do depends on the dominant scientific paradigm (of which definitions are an integral part) that shapes and directs the choice of research problems. It will never be value free.

The historian L. White (1967) thinks we are unquestionably inheritors of the Baconian creed that scientific knowledge leads to technological power over nature and that power is good. Most weed scientists subscribe to the Baconian creed without thinking about it or considering how it determines where we are going. Modern, successful science and technology are Western and weed science, especially as embodied in herbicides, is part and parcel of Western modernity. White suggests we have moved from man-as-part-of-nature to man-over-against nature. Humans stand in sharp contrast to all other species which must adapt to the environment. Man and nature have become separate. Man is master now, and it was meant to be so. Man's power yields dominion and the ability to subdue nature even though we strive to obtain goals that are ecologically unsound and unsustainable. Nature and natural things are judged by what they can do for man, not by any value judgment about intrinsic natural patterns that control us and are affected by our actions. The basic character of modern man—we are programmed to desire, not to appreciate, is one source of the problem (Ridley 2010, p. 27). Our lives include endless pursuit of things, exploitation and despoliation of nature and exploitation of other people. Thus, we are alienated from nature and willingly accept definitions which create attitudes that make some things undesirable because they do not fit our immediate goals. White doubts that applying more science and more technology to problems created by science and technology will avoid a disastrous ecological backlash. He wants us to rethink the Christian axiom that nature has no reason for existence except to serve man.

Black (1970) agrees with White that there is an ecological crisis, but sees a different cause. He thinks our ecological crisis does not stem from our Judao-Christian heritage to subdue and dominate, as White does, but rather from a fundamental uncertainty about the past and future of man. Such doubts and the questions they create are often explored in myths (e.g., Prometheus). Every individual must think about and try to answer the eternal questions addressed by some myths and religions: Who am I? What is the meaning of my life? What is the purpose of existence? Whether we confront, ignore, or remain unaware of these questions, they cannot be rejected as irrelevant. We, those who have preceded us, and those who follow will still struggle with them. Black concludes "that almost the only course open to western man is based on a vision of mankind stretched out along the dimension of time. By redefining mankind in terms of the whole of humanity dead, living, or as yet unborn, we may perhaps be able to assess what we do in terms of the good of mankind, regardless of the position of the individual along the time axis of the world". Black sounds much like Heilbroner (1975) who argues for the "transcendent importance of posterity" to each individual.

Wendell Berry (1981) disagrees with White's interpretation of Genesis and the Christian message, but agrees with White's hypothesis that we have regarded ourselves, especially in agriculture, to be in a state of war with nature. Berry suggests we must not use the world as if we created it. Land and earth exist independent of people, and their existence is not dependent on human purpose. White wants us to rethink our reason for being, and Black challenges us to include a future dimension to life while redefining ourselves in relation to others. Berry fundamentally agrees but sees that "the most necessary thing in agriculture is not to invent new technologies or methods, not to achieve breakthroughs, but to determine what tools and methods are appropriate to specific people, places, and needs, and then to apply them correctly".

Marx (1970) cites White and agrees that Christianity has given the West an "aggressive man-centered" environmental attitude; an attitude that assumes that only when man achieves the ultimate, hoped for, unity with God can we transcend nature and achieve dominance and adjust or force nature to our ends which, of course, are what God intended from the beginning. Marx suggests that even conservationists see a world that exists apart from and for the benefit of mankind. The philosophical root of Marx's ecological ideal is the secular idea that man is "wholly and ineluctably" embedded in the natural world. Kass (1999 p. xx, xii) agrees and thinks that at the end of the twentieth century we are befuddled about how to act and what to think about our relationship to the natural world. In his view the cause of our befuddlement is spiritual distress and alienation from the natural world caused primarily, but not only, by the dominant scientific view of the world. The root of the problem for Kass and Marx is philosophical.

Certainly it is the view of many that our agricultural-ecological problem is an attempt to transcend nature and it is rooted in the West's profit-oriented, business dominated society (Marx). This results in violation of ecological standards for short-term gain, but no obvious (short-term) cost; a cost that is highly likely to be exter-nalized.[6] Marx cites the omnipresent expansionary or frontier ethos of the US which

---

[6] An externality is a cost that is not reflected in price or more technically, a cost or benefit for which no market mechanism exists. In the accounting sense, it is a cost that a decision maker does not have to bear, or a benefit that cannot be captured. From a self-interested view, an externality is a secondary cost or benefit that does not affect the decision maker.

"like a powerful, ideological hormone, stimulates reckless, uncontrolled growth of each cell in the social organism." For Marx, the fundamental confrontation is the extreme imbalance between each citizen's desire to satisfy the growing total of individual wants and the limited capacity of a closed ecosystem to give everyone what they want. Some question if the 9 + billion people projected to be on earth in 2050 will be able to have what they need, plus what they may learn to want. The capacity of agricultural scientists, indeed the human capacity, for denial is clear. We are optimists. What could be more human than believing that everything will be all right, that the problems can be solved, that in spite of clear evidence of harm (dying oceans, polluted water, diminishing water supply, eroding soil, species loss, global warming, etc.) we can feed the world (Ridley 2010, 121–156)?

Is it possible given our unexamined view of science and the world (we see things as we are) that our definition of simple things can lead to trouble and misdirect our power? I suggest that weed science is a discipline that has lost its way because of wholesale acceptance of unexamined definitions and technologies. It is my view that it will be helpful if we study our history, not assume it, and evaluate it as a guide to the future. Scientists know, as Mortenson (2009, p. 21) reminds us, that hope resides in the future, while perspective and wisdom are almost always found by looking to the past.

The History of Weed Science

Weed Scientists, in fact most scientists, are not historians. If weed scientists think about the history of their work, they probably assume it is similar to the history of similar groups. I suggest that this assumption of similarity is an error. I do not purport to be an historian, but after 45 years experience, I can comment as an internal historian. Few weed scientists have investigated the history of weed science. The historical accounts which have been written are accurate, but are almost exclusively chronologies of the development of chemical weed control (Crafts 1960; Smith and Secoy 1975, 1976; Timmons 1970; Upchurch 1969; Zimdahl 1983), a worthy topic. A more complete historical study can be found in Zimdahl 2010—a History of Weed Science in the United States. A summary of selected excerpts follow.

Upchurch (1969) says that "one approach to studying the beginning of weed control is to examine the appearance of various herbicides". He is correct and it is interesting that his approach is the one commonly taken by writers. Many writers cite the early work of Bolley in North Dakota in 1908, and the nearly concurrent work of Frenchmen Bonnet, Martin and Duclos and Germany's Schultz (cited in Crafts and Robbins 1962). Each of these men used solutions of copper salts for selective weed control in small grains; later iron sulphate and sulfuric acid were used. Succeeding work in Europe observed the selective herbicidal effects of metallic salt solutions or acids in cereal crops. Names encountered frequently include Rabate in France (Rabate 1911), Morettini in Italy (1915), and Korsmo in Norway (1932). More recent historical accounts almost always cite the first synthesis of 2,4-D by Pokorny in 1941 (cited in Klingman et al., 1982 and in many textbooks), the discovery of its growth regulating properties (Zimmerman and Hitchcock 1942) and the first reports of its herbicidal activity in the field (Hamner and Tukey 1944a, b; Marth and Mitchell 1946). At the same time work was progressing in England on herbicidal uses of MCPA, a close relative of 2,4-D (Slade et al. 1945). The first paper on a non-phenoxy acid derivative useful as a herbicide was by Bucha and Todd (1951) who reported the herbicidal properties of monuron the first of many phenyl urea herbicides. When a weed science textbook or paper discusses history, they accurately claim that weeds have been with us since settled agriculture began. Smith and Secoy (1976) state that harmful effects of weeds were known to early historic man and cite the evidence of hoes and grubbing implements that have been found. They report that nearly all books from Theophrastus to modern times have mentioned weeds and their detrimental effects. It is not hard to find examples of man's battle with weeds. The thistles and thorns of Genesis (3:17–18), the parables of the sower (Matthew 13:18–23), and the tares (Matthew 13:25–30) are among the earliest.

In many parts of the world farmers are people with hoes. It is often not noted, those who use the hoes are women. Weeds, in much of the world, are part of the burden of agriculture. Control, if achieved, is an incidental part of production. It has progressed from hand removal, primitive hand tools, animal and tractor powered implements to herbicides. Weed science and planned weed control began following World War II as a result of the discovery of the herbicidal properties of 2,4-D. This relatively (65 years) new science of weeds is and has been from its inception, a pragmatic science. Weed scientists are problem solvers.

There were herbicides before 2,4-D, but none were as cheap, as effective, or as selective. Crafts (1960) traces the history of herbicide development from the early workers, cited above, to the discovery in the 1920s of the bindweed control potential of a dilute solution of sodium arsenite and its apparent translocation in plants. In Colorado, workers applied carbon bisulfide as a soil fumigant to control *Phylloxera* (an insect family related to aphids) which feeds on roots of grapes (*Vitis vinifera*) often with devastating effects. They noted bindweed was killed and soil fumigation with carbon disulfide came into wide spread use.

Crafts (1960) presents the early work on sodium chlorate, sodium arsenite and sodium dinitrocresylate. These true and interesting accounts confirm that the history of weed science has been dominated by the development and evolution of chemical control, although other common methods are acknowledged. The dominance of herbicides in weed science meetings and journals is still true as shown in Tables 1 and 2. The emphasis on herbicides in Weed Science has decreased whereas it has increased in Weed Technology, which is appropriate given the intended purpose of each journal. One must note that the percentage (38%) of papers devoted to weed biology and ecology from 2005 to 2010 is nearly equal to the percentage devoted to herbicides (34%). Papers on biology/ecology of weeds increased from 14.3% prior to 1970 to 23.5% in the 1990s (Abernathy and Bridges 1994). In the same study 55,000 references on weeds and their control were found in the Weed Abstracts section of CAB abstracts. They were divided

Table 1 The percentage of papers focused on herbicides at weed science meetings attended by the author

Society	Year	Percetage of Papers on Herbicides
Western Society of Weed Science	1992	55.5
	1993	62.5
	1995	47.2
	1996	46.8
	1998	54.0
	1999	58.5
	2000	59.0
	2001	59.8
	2002	62.0
	2003	53.5
	2004	50.5
	2005	56.3
	2006	43.0
	Average	54.5
Weed Science Society of America	1987	61.9
	1989	60.4
	1990	63.6
	1991	65.9
	1993	62.6
	1994	62.6
	1995	52.5
	1996	55.0
	1997	47.0
	1998	45.0
	1999	55.0
	2000	50.0
	2001	49.8
	2003	46.0
	2006	48.0
	Average	55.0

into eight categories. Based on these data Abernathy and Bridges disagreed with the claim that emphasis on herbicides had not changed over several decades. Their conclusion is based on the percentage of papers in the category herbicide activity/selectivity/efficacy—57% of all papers prior to 1970. That declined to 29% in the 1990s. They claimed this indicated declining emphasis on herbicides. However, if one makes the reasonable assumption that four of their categories (Environmental fate/transport, Herbicide metabolism/mode of action/physiology/biochemistry, Application technology, and Herbicide resistance deal primarily with herbicides and combine the data from these categories with the category above, those that deal with herbicides from the pre-1970 group through the 1990s are 71% of the total. This supports the claim that up to 1994, weed science research was dominated by the development and evolution of chemical control. It is important to note that other journals (e.g., Bioinvasions, Bioscience, Ecology, Plant Ecology)

Table 2 Percentage of
papers focused on herbicides
over the most recent 6 years
in the journals Weed Science,
Weed Technology, and
Invasive Plant Science and
Management

Journal	Year	Percentage of Papers on Herbicides
Weed Science	2005	38
	2006	25
	2007	32
	2008	36
	2009	38
	2010 (3 issues)	44
	Average	35
Weed Technology	2005	84
	2006	55
	2007	66
	2008	67
	2009	69
	2010 (3 issues)	62
	Average	67
Invasive Plant Science and Management (First published in 2008)	2008	26
	2009	21
	Average	23.5

The Weed Science Society of America publishes all three journals

publish articles that deal with weeds, especially invasive species. The articles in these, and other journals, emphasize study of life cycles and ecological interactions of weeds rather than control, which one hopes will lead to better control.

The need for weed control has long been acknowledged, but the introduction of herbicides, which enabled selective control of weeds, was the stimulus for organization and development of weed science. Timmons (1970) noted that state weed laws were first enacted between 1721 and 1766, but they were not weed laws as we now know them. The laws were directed at control of plant diseases spread by weeds which often serve as alternate hosts. There were only a few Extension Service publications on any aspect of weeds between 1860 and 1900 (Timmons). Few others were added until the 1950s. Weed control was a minor part of agronomy, botany, horticulture, and plant physiology until the 1950s (Timmons). The journal Weeds (now Weed Science) was founded in 1951. The Weed Science Society of America, publisher of the journal, was not founded until 1956. The primary European weed journal, Weed Research, was founded in 1961.

The brief history weed science has been suffused with what Veblen (cited by Galbraith, 1981) calls exoteric knowledge. It is knowledge that has negligible academic prestige, but is very useful. It is pragmatic, problem solving knowledge. Weed science is dominated by pragmatism. No one has or is likely to accuse weed scientists of generating only esoteric knowledge which Veblen said has great academic prestige, but is without economic or industrial utility.

The History of Other Plant Protection Disciplines

A more complete study of the history of other plant protection disciplines can be found in Zimdahl (2010). A brief presentation follows.

The history of weed science is brief compared to that of entomology and plant pathology. The latter two began as applied sciences, and both continue significant applied research. They differ from weed science in that each has causal organisms that in many cases have been isolated and identified. The effect of the organism in plant pathology is called a disease, whereas insects cause growth malformations or reduced growth. The cause is usually apparent when a disease or insect is present. Weed presence is obvious, but the yield reducing effect of low densities on the crop is not equally obvious. Weeds are often just there and the crop seems all right because weeds do not cause the obvious effects that other organisms do. Weeds don't eat things or immediately cause wilt or malformations.

All three plant protection disciplines trace their origin to the Bible and cite it as evidence of ancient awareness of problems caused by the pests of most importance to them. Christian literature is replete with examples of insect scourges, and blast and mildews are common in the Old Testament. The first to write about plant diseases of trees, cereals and legumes was Theophrastus (370–286 BC) (Agrios 1979). His approach was observational and speculative rather than experimental and little more was added to the knowledge of plant pathology for the next 2000 years.

In the 16th century, botanists were intent on naming and describing plants rather than investigating how they grew and developed. The dogma of constancy of species led to the apparently logical assumption that fungi arose by spontaneous generation. Farmers of that age accepted disease as an inevitable concomitant of crop growth just as unfavorable soil and climate were conditions agriculture had to accept. From the sixteenth well into the eighteenth century, many prominent botanists were skeptical of spontaneous generation, but the theory held (Agrios 1979). Discovery of the compound microscope in the seventeenth century made new observations possible and began a new phase of plant pathology. Working with a microscope in 1675, Anton van Leeuwenhoek discovered bacteria, but his discovery did not influence spontaneous generation. Fungi were a result not a cause of disease (Agrios). In the nineteenth century the influence of human pathology on interpretation of plant disease was strong. A comprehensive paper on fungi was written by the German botanist Anton de Bary (1831–1888) when he was 22 years old and he thereby entered the controversy about the relation of fungi to plant disease. He opposed spontaneous generation and claimed that fungi were not symptoms, but parasites that caused diseases.

Science, in the 1800s, was shifting from a philosophical approach to experimentation and inductive reasoning. However, throughout the eighteenth century, the prevailing concept remained that lower organisms arose *de novo* from inanimate substrates. Spallanzani, at the end of the century, and Tyndall and Pasteur several years later, finally disproved spontaneous generation (Tarr 1972). Even after 1860, when Pasteur showed that microorganisms arose from pre-existing organisms and that fermentation was a biological phenomenon rather than a purely chemical one, spontaneous generation persisted as a theory. Tradition dies hard, even in objective science. Pasteur's work was not widely accepted and the germ theory of disease for man and animals was not established in 1860. De Bary's continued work provided a foundation for the germ theory of plant disease, and he is regarded as the father of modern plant pathology. Plant pathology texts first appeared in Europe after the mid-1850s and took a taxonomic approach (Tarr 1972). T.J. Burrill of the University of Illinois (Walker 1969) was the first to relate a bacterium to the cause of fire blight of pears. Leadership in the study of bacterial diseases was by American scientists in the latter part of the nineteenth century. Burrill's work was greeted with skepticism by Europeans who regarded themselves as leaders.

The US Department of Agriculture was founded in 1862 and the Division of Botany was organized in 1885. The section on mycology was begun in 1886. Its name was changed to the section of vegetable pathology in 1887 and to the Division of Vegetable Pathology in 1891 and finally to the Bureau of Plant Industry in 1901. The Federal Hatch act, which established state land grant agricultural experiment stations, was passed in 1887 and by 1890, plant disease studies had begun in many states. In fact, Bolley (Zimdahl 2010), who is properly cited by weed scientists as a founder of selective chemical sprays for weeds was a plant pathologist in the North Dakota agricultural experiment station.

Plant pathology began to emerge as a discipline toward the end of the nineteenth century and was well established in the US and Europe by the beginning of the twentieth century. The emerging discipline was characterized by studies of taxonomy and description of diseases and their causal agents (Tarr 1972). Early work did not center on disease control. It is not known if control was even a goal of pathologists. When potato blight swept across Europe in the 1840s and the Irish potato famine occurred, blight's cause was not agreed upon. Only slowly was it accepted that mold associated with affected potatoes was the cause and not the result of disease (Tarr). De Bary established that *Phytophthora infestans* was the cause of late blight of potatoes. However, it is important to note that the research was directed at finding the cause not a cure for an undefined problem.

Because man has always been more concerned with human and animal health than plant health and because insect vectored diseases became the province of entomology, plant pathology was slow to develop. Human and animal pathology dealt with diseases whose importance was accepted, but plant pathology has never had the comprehensive scope that medical or veterinary science has enjoyed.

We know that fungicides were developed slowly through empirical observation before the germ theory of disease was well established. Sulfur was used for many years as a fungicide. A landmark was its introduction in 1802 by William Forsyth (1737–1804), who, as gardener to King George III, used lime sulfur for control of mildew on fruit trees (Tarr 1972). Tarr regards the turning point in use of chemicals for plant disease control as the introduction by Millardet (1838–1902) in about 1885 of Bordeaux mixture (copper sulfate, lime, and water) in France for control of downy mildew *(Plasmopara viticola)* of grapes. It is also cited as one of the starting points for herbicides because an unknown good observer noted that it turned leaves of some species of mustard black. It acted as a selective herbicide and the event is often cited as the origin of work on herbicidal properties of metallic salts. Organic fungicides began appearing in 1913 when chlorophenol of mercury was used to treat wheat seed for control of bunt or stinking smut *(Tilletia caries* or *T. foetida)*. It was followed by red copper oxide and zinc oxide.

Entomology's history is similar to that of plant pathology. The science developed early and Richard (1973) provides a continuous curve showing the life history of 56 entomologists from 1700 to 1950. Much development occurred in China in the fifteenth and sixteenth centuries and earlier (Konishi and Ito 1973). But China was an inward looking society with no firm external linkages, and little development occurred. The basic character of Chinese scientific philosophy was organismic or holistic, not experiential and analytical. The Chinese saw the world as a pattern of relationships which were to be studied and understood.

In the West, zoology did not exist until Aristotle and the term "insect" is not found in Biblical Hebrew or other languages of the era (Harpaz 1973), but we know that people were not ignorant of insects. Entomology was a systematized and descriptive science in the West from 1700 to the early 1800s. The entomologist's task was to describe and understand insects, their life cycles, hosts, and the damage they did. Probably because they could not, their primary task was not to control or devise control programs for insects. The first part of the 19th century saw formation of the most important European entomological societies: France, 1832; UK, 1833; Germany and Holland, 1857; and the Pennsylvania society in the US in 1842 (Smith et al. 1973).

An Historical Conclusion

Weed science cannot claim the historical lineage that either of the other two major plant protection disciplines can. Although weeds have been around as long as insects and plant diseases, they have not been studied as long. Although it is reasonable to argue that weed control has been practiced since people began to grow crops. The major actors in the history of weed science all completed their education and developed their careers in the twentieth century. Included among them are six major figures (a more complete biography of 25 weed scientists is included in Chap. 2 of Zimdahl 2010).

Wilfred W. Robbins was born in Mendon, Ohio May 11, 1884 (died 1952). He received his Bachelor's and Master's degrees from the University of Colorado and a Ph.D. in Botany from the University of Chicago in 1917. The early part of his career was spent as an instructor in botany and forestry and as a botanist in the agricultural experiment station at Colorado Agricultural College, Fort Collins. He moved to the University of California at Davis in 1922 where he was chairman of the Botany division of the College of Agriculture for 29 years. There he began a program on weed control and developed the first classroom instruction about weeds (Crafts and Robbins 1962). He wrote one of the first textbooks on weed control and was instrumental in establishing the first US weed society—The Western Weed Control Conference in 1938.

James W. Zahnley (1884–1975) was born near Dwight, Kansas, received Bachelor's and Master's degrees from Kansas State University, and spent his entire career in Kansas. He joined the agronomy faculty at Kansas State University in 1915 where he pioneered chemical weed control investigations. He discovered the use of sodium chlorate for control of field bindweed (*Convolvulus arvensis* L.) and Russian knapweed (*Acroptilon repens* (L.) DC.) and was involved with early experiments on the herbicidal potential of sodium trichloroacetate for control of perennial grasses.

Many of those who retired from active weed science careers in the latter part of the twentieth century received their education under the guidance of Charles J. Willard, Professor of Agronomy at The Ohio State University and a founder of the Weed Science Society of America. Willard was born in Manhattan, Kansas in 1889 (died 1974). He received his education at Kansas State, the University of Illinois and Ohio State where he remained on the faculty for 42 years. He began the weed control program at Ohio State with studies of chemical control in 1927. He served as major professor for many men who went on to build weed science after World War II.

F. L. "Tim" Timmons was born in Little River, Kansas in 1905 and received a Bachelor's and Master's from Kansas State University and his doctorate from the University of Wyoming. His professional career was spent in the West and included work in Kansas, Utah, and Wyoming. When he retired, he was recognized as an expert on range land weed control and weed control in aquatic systems.

Erhardt P. "Dutch" Sylwester (1906–1975) was Professor of Botany and Plant Pathology at Iowa State University. For those who knew him, he remains the epitome of the extension weed specialist, which he was for 30 years. His skill was a combination of his knowledge of Iowa's crops and weeds and his ability to teach weed control to farmers whose questions come from experience and need. Farmers demanded answers immediately not in the next day's lecture. Dutch was one of the first to receive a sample of the "magic" herbicide 2,4-D in 1945. He had been preaching the importance of weed control through tillage, clean seed, and good farming practices for many years and with 2,4-D he had a tool to accomplish selective weed control in corn, Iowa's major crop. He developed weed control with herbicides in Iowa, and his program became a model for much of the corn belt.

Another pioneer in weed science was Kenneth P. Buchholtz (1915–1969) who received his undergraduate degree from Washington State College in 1938. He went on to receive a Master of Science and Doctorate from the University of Wisconsin where he remained on the faculty in Agronomy and began the program in weed science. He was a pioneer whose career paralleled the development and use of selective herbicides. Many of the twentieth century's active weed scientists received their education in his program at Wisconsin.

Finally, one must include Alden S. Crafts (1897–1990), a Colorado native who spent his professional career at the University of California at Davis. He began as a colleague of W. W. Robbins and built on the foundation Robbins laid for weed work in California. He received his doctorate in plant physiology from California and did much of the pioneering work on mode-of-action of herbicides and laid the basis for modern studies through his work with phloem structure and function, water relations of plant cells and autoradiography.

The point to be drawn from these brief biographies is that these were all men who completed their education and developed their careers in the twentieth century. They began weed science but were not educated as weed scientists because there were no weed scientists who preceded them. They founded a modern science and helped develop a scientific society (Weed Science Society of America) whose membership peaked at nearly 3,500 in 1994. Membership and attendance at the Society's annual meeting have declined since 1994. In 2010 the society had fewer than 1,300 members.[7] The decline should not be interpreted as a decrease in the need for weed control. It is primarily due to consolidation in the chemical industry and reduction in staff of weed science programs in US universities.

One mark of a mature profession is awareness and understanding of its own history (White 1968). When we know our history then, and only then, can we judge what it means and who we are. History cannot be assumed any more than the results of an experiment can. The history of weed science is different from that of other major plant protection disciplines, and I suggest that the science has been shaped by that history although, in general weed scientists are not aware of its influence.

As we have seen, entomology and plant pathology began before their practitioners had the ability to control causal organisms. Early scientists studied the causal organism, its life cycle and how it interacted with a host. It was their only choice because effective control was not possible. My hypothesis is that weed science missed this foundation-forming phase so prominent in the history of other plant protection disciplines. Prior to World War II, weed scientists were few in number and control dominated their thinking, even though their ability to control was limited. During their formative years, other plant protection disciplines were compelled to develop an understanding of the organisms they can now control. Weed science never went through this phase because those who began to try to solve weed problems were so closely followed by others to whom technology had provided the means of control. When herbicides were developed after World War II, weed science began to develop because the ability to control was at hand. The early scientists did not need to develop understanding of a weed's biology or ecology. Their task was to figure out how to use herbicides to kill weeds selectively in crops. It was a demanding task and it remains a difficult job. Great successes have been achieved and much remains to be done to perfect chemical technology. However,

---

[7] Personal communication, J. Lancaster, Weed Science Soc. of America, Lawrence, KS, November, 2010.

much of it can be done without complete, and often with only limited understanding of the weed to be controlled. It must be pointed out that all plant protection disciplines were equally eager to join the pesticide parade after World War II.

Weed science and, in large measure, other pest control disciplines, are now chemically-based, control-oriented, scientific endeavors that often do not rely on understanding of the pest's habits or habitat prior to development of control programs. When a new weed appears, many weed scientists look for an herbicide or ask what new one is coming along to solve the problem. Questions about what caused the problem or alternative control techniques do not arise as frequently. For many years weed scientists largely ignored study of weed ecology, but based on the orientation of journal papers presented above, it is clear that this is changing. However, a great deal of effort and time of talented people remains focused on control, often with a herbicide. Vegetation management strategies can be distorted by undo emphasis on herbicides that offer quick, predictable success. Herbicide intensive strategies frequently neglect or ignore other management techniques.

Weed science has been and is confronted with several serious challenges including increasing resistance of weeds to herbicides, questions about environmental effects, problems concerning human and mammalian safety, emergence of new weed problems, and increasing regulatory restraints on herbicide development. These challenges have drawn a creative response and a perceptible shift from strict chemical weed control to the developing concept of weed management. Weed management moves away from strict reliance on control of an existing problem and gives greater emphasis to prevention of seed and vegetative propagule production, reduction of weed emergence in a crop and minimization of competition during a crop's life (Aldrich 1984). Weed management emphasizes integration of techniques to manage or anticipate problems before they occur rather than solving them after they are present. Weed management will not eliminate the need for control nor does it advocate that today's best techniques be abandoned in favor of a return to some pre-existing, pristine kind of agriculture where a mythical balance of nature was preserved. It can maximize crop production where appropriate and attempt to optimize grower profit by integrating preventive techniques, scientific knowledge, and managerial talent. Weed scientists need and are working to develop additional information in all of these areas. An important part of the task is to develop knowledge of weed biology and ecology so we understand what makes a weed a weed and then use appropriate management tools rather than prophylactic measures that often result in amazing short-run solutions, but may worsen long-term problems.

However, we still practice an agriculture and a weed science that have great (sometimes complete) reliance on herbicides for weed control. Berry (1981) and White (1967) have argued that this kind of reliance is related to the way each of us views the world—our definitions—and regards history as irrelevant. The time has come to rethink the fundamental goals of the science. Scientists, including weed scientists, commonly think they can solve the problems science and technology have created by applying more of the same kind of science and technology. In weed science, it appears to have worked, so far. But, one must consider if it really has and will continue to work. This requires knowing and evaluating, in a comparative way, all positive and negative effects. Well-known positive effects include increased crop yield, and grower profit. The value of each is determined by economic analysis. Do the benefits exceed the costs? Negative effects (e.g., soil, water and air pollution; pest resistance; bioaccumulation; effects on biodiversity, ecosystem stability, and other species) are also well-known and, while unintentional, are more difficult to quantify and evaluate in terms equivalent to those used to assess benefits.

Any science advances not by authenticating everyday experience, but by grasping paradox and adventuring into the unknown (Boorstin 1983). Weed scientists believe correctly, with a great deal of supporting evidence, that their science has contributed to increasing the world's food supply and reducing the human drudgery of weeding—one of humanity's most onerous and time consuming tasks. But weed scientists must face the paradox that reliance on one major means of achieving a desirable end may be counter productive. Weed science should not abandon what is known, and the useful management techniques derived therefrom. Rather there should be a continuing, rigorous examination of the science's basic precepts, leading toward appropriate changes.

Do you see any clue?
    You have furnished me with seven but of course I must test them before I can pronounce upon their value.
    You suspect someone?
    I suspect myself.
    What!
    Of coming to conclusions too rapidly. (Doyle 1927)

References

Abernathy JR, Bridges DC (1994) Research priority dynamics in weed science. Weed Technol 8:396–399

Agrios GN (1979) Plant Pathology. Academic Press, New York, pp 9–13

Aldrich RJ (1984) Weed-crop ecology: principles in weed management. Breton Publishers, North Scituate, pp 5–6

Anonymous (1983) Herbicide Handbook of the Weed Science Society of America, 5th edn. Weed Sci Soc Am Champaign, IL, p. xxiv

Bailey LH, Bailey EZ (1941) Hortus the Second. MacMillan, New York, p 778

Baker HG (1965) Characteristics and modes of origin of weeds. In: Baker HG (ed) Genetics of colonizing species. Academic Press, New York, pp 147–172

Berry W (1981) The gift of good land: further essays cultural and agricultural. North Point Press, San Francisco, p 280

Black JN (1970) The Dominion of man: the search for ecological responsibility. Edinburgh University Press, Edinburgh, p 169

Blatchley WS (1912) The Indiana weed book. Nature publishing company, Indianapolis, p 191

Bolley HL (1908) Weeds and methods of eradication and weed control by means of chemical sprays. N Dakota Agric College Expt Station Bul No 80:511–574

Boorstin DJ (1983) The discoverers: a history of man's search to know his world and himself. Intro to Book 3–Nature, Random House, p 291

Brenchley WE (1920) Weeds of farm land. Longmans Green & Co., Inc, New York, p 239

Bucha HC, Todd CW (1951) 3(p-chlorophenyl)-1, 1-dimethylurea -A new herbicide. Science 114:493–494

Bunting AH (1960) Some reflections on the ecology of weeds. In: Harper JL (ed) The biology of weeds. Blackwell Science Publications, Oxford, pp 11–26

Capra F (1982) The turning point. Bantam Books, New York, pp 87, 252–253

Covey SR (1989) The 7 habits of highly effective people: powerful lessons in personal change. A Fireside Book. Simon & Schuster, New York, p 358

Crafts AS (1960) Weed control research—past. present and future. Weeds 8:535–540

Crafts AS, Robbins WW (1962) Weed control: a textbook and manual, 3rd edn. McGraw-Hill Book Co., Inc, New York vii–vii and 173

Doyle SirAC (1927) The complete sherlock holmes, vol I. Doubleday and Co. Inc., Garden City, p 456

Eiseley L (1971) The night country. C. Scribner's Sons, New York, p 240

Emerson RW (1876) Fortune of the epublic. In: Miscellanies. The complete works of Ralph Waldo Emerson, vol Xl, Houghton Mifflin, NY, pp 509–544

Freedman, DH (2010) Lies, dammed lies, and medical science. The Atlantic, November, pp 76–78, 80–82, 84–86

Galbraith JK (1981) A life in our times. Houghton-Mifflin Co., Boston, p 24

Glass B (1965) The ethical basis of science. Science 150:1254–1261. Also pp 43–55 In: Bulger RE, Heitman E, Reiser SJ (1993) The ethical dimensions of the biological sciences. Cambridge Univesity Press, Cambridge

Hamner CL, Tukey HB (1944a) The herbicidal action of 2, 4-dichlorphenoxyacetic acid and 2, 4, 5-trichlorophenoxyacetic acid on bindweed. Science 100:154–155

Hamner CL, Tukey HB (1944b) Selective herbicidal action of midsummer and fall applications of 2, 4-dichlorophenoxy acid. Bol Gaz 106:232–245

Harpaz I (1973) Early entomology in the Middle East. In: Smith RF, Mittler TE, Smith CN (eds) History of entomology. Ann Rev Inc., pp 21–36

Heilbroner, RL (1975) What has posterity ever done for me? New York Times Magazine, pp 14–15, Jan 19

Kass LR (1999) The Hungry soul: eating and the perfecting of our nature. University of Chicago press, Chicago, p 248

Kirschenmann F (2010) Imagining resilience. The Leopold letter 22(3):5

Klingman GC, Ashton FM, Noordhoff L (1982) Weed science: principles and practices. Wiley, New York, p 12

Knusli E (1970) History of the development of triazine herbicides. Residue Rev 32:1–9

Konishi M, Ito Y (1973) Early entomology in East Asia. In: Smith RF, Mittler TE, Smith CN (eds) History of Entomology. Ann Rev Inc, pp 1–20

Korsmo E (1932) Undersokelser 1916–1923. Over ugressets skadevirkninger og dets bekjempelse. Akerbrucket. Johnson and Neilsens Boktrykkeri, Oslo, 411 pp

Little W, Fowler HW, Coulson J (1973) The shorter Oxford english dictionary on historical principles. In: Onions CT (ed) Etymologies revised by GWS Friedrichsen, vol 2, 3rd edn. Clarendon Press, U K, p 2672

Marth PC, Mitchell JW (1946) 2, 4-dichlorophenoxyacetic acid as a differential herbicide. Botan Gaz 106:224–232

Marx L (1970) American institutions and ecological ideals. Science 170:945–952

Morettini A (1915) L'impiego dell'acido solforico per combattere lecrbe infeste nel frumento. Staz Sper Agr Ita! 48:693–716

Mortenson G (2009) Stones into schools—promoting peace with books not bombs in Afghanistan and Pakistan. Penguin books Inc, New York, p 420

Muenscher WC (1960) Weeds, 2nd edn. MacMillan Co., New York, p 560

Rabate E (1911) Destruction des revenelles par l'acide sulfurique. J d'Agr Prat (n.s.21) 75:497–509

Richard G (1973) The historical development of nineteenth and twentieth century studies on the behavior of insects. In: Smith RF, Mittler TE, Smith CN (eds) History of Entomology. Ann Rev Inc, pp 477–502

Ridley M (2010) The rational optimist: how prosperity evolves. Harper Collins publishers, New York, p 438

Ruttan V (1986) Increasing productivity and efficiency in agriculture. Science 231:781

Slade RE, Templeman WG, Sexton WA (1945) Plant growth substances as selective weed killers: differential effect of plant-growth substances on plant species. Nature (Long) 155:497–498

Smith AE, Secoy DM (1975) Forerunners of pesticides in classical Greece and Rome. J Agric Food Chem 23:1050–1055

Smith AE, Secoy DM (1976) Early chemical control of weeds in Europe. Weed Sci 24:594–597

Smith RF, Mittler TE, Smith CN (1973) History of Entomology. Ann Rev Inc, 517 pp

Tarr SAJ (1972) Principles of Plant Pathology. Winchester Press, NY, pp 8–17

Thomas WL, Jr (ed) (1956) Man's role in changing the face of the Earth. An international symposium under the chairmanship of Sauer C, Bates M, Mumford L. Sponsored by the Wenner-Gren Foundation for Anthropological Research. University of Chicago Press, Chicago, 1236 pp

Timmons FL (1970) A history of weed control in the United States and Canada. Weed Sci 18:294–306

Upchurch RP (1969) The evolution of weed control as a science. Indian J Weed Sci 1:77–83

Walker JC (1969) Plant Pathology. McGraw-Hill Book Co, NY, pp 14–46

White L Jr (1967) The historical roots of our ecological crisis. Science 155:1203–1207

White L Jr (1968) The dynamo and the virgin reconsidered: essays in the dynamism of western culture. The MIT Press, Cambridge, p 186

Zimdahl RL (1983) Weed science—a brief historical perspective. Weeds Today 14(1):10–11

Zimdahl, RL (2010) A history of weed science in the United States. Elsevier, Inc, Burlington, 207 pp

Zimmerman, PW Hitchcock AE (1942) Substituted phenoxy and benzoic acid growth substances and the relation of structure to physiological activity. Contributions. Boyce Thompson Institute 12:321–343

Chapter 2
Pesticides and Value Questions

Unexamined, rapid conclusions about anything can be serious errors of judgment, and actions based on them may lead to unanticipated, perhaps undesirable consequences. Pesticide technology has not lacked challenges to conclusions about its role and value. These have often been met by pleas for scientific objectivity and dismissal of challenges as emotionally laden and lacking in understanding of the necessity of high agricultural production, pesticide's role in maintaining production, and the extensive safety evaluation mandated before a pesticide ever reaches the US market. While not without foundation, these pleas do not allow careful consideration of other arguments and points of view and their logical conclusions. I first encountered the problem of reasoning with opposing points of view in the late 1960s and early 1970s during the 2,4,5-T controversy. My thoughts were clarified as I struggled with the inevitable value questions (Zimdahl 1972). They are reproduced in a slightly modified form below.

> Once upon a time all life seemed to be in harmony with its surroundings. Once upon a time we lived in an orderly world in which change occurred so benevolently it was called progress. There was a place and a time for everything. Six days for work and one for being told the meaning of work. The professor and the pastor spoke without hesitation. Dad and Mom told us what good boys and girls did (or mostly didn't) do. The newspaper told it like it was. Uncle Sam could be trusted to make the world safe for democracy. And filling in all the cracks between constituted authorities were reason and common sense.
>
> Keen (1969)

Then something happened. Something in the pesticide industry that questioned established reason and common sense. The nagging problem that the ultimate effects, if any exist, of long-term low-level exposure to pesticides had not been well enough explained to answer some fundamental value questions. The public's

---

Adapted with permission from the Bulletin of the Entomological Society of America. Zimdahl, R. L. June: 109–110, 1972.

willingness to ask more difficult technical questions became apparent. Often the answers were unknown and had not even been measurable in the early years of pesticide development. Most embarrassing, they were questions that those who developed, recommended and used pesticides had not even thought of asking. In the 1990s the pesticide industry and pesticide users were challenged for answers about long-term induction of cancer or birth defects, alteration of genes, and interaction with other chemicals. Agricultural scientists who knew they were helping to produce more food to feed a hungry world by changing (improving) production practices to more efficiently manage insect, disease and weed problems were not ready for the criticism heaped on them for pursuing such a universally acceptable goal. In attempting to sort out my thoughts on the questions raised about chlorinated hydrocarbons, 2,4,5-T and organophosphate insecticides, I developed an awareness of a level of questioning that had gained public credibility. Although it was not entirely in harmony with my own, it was important.

The Value Issue

There is a general agreement among men that all ends or goals should be good. In a value question, problems arise because people do not agree on what is good or what is true and therefore on what ought to be done. There is disagreement concerning use of pesticides as a means to achieve the desirable goals of increased food production and improved public health. Ends may be analyzed to determine their value. But those who work with pesticides are compelled to analyze them as means to an end and to determine the compatibility of means and ends. Ends pre-exist in means. Pesticides contain only natural ends not the ends predicted by ardent advocates or determined opponents of their use. As a seeker of truth concerning untoward effects of these pesticides, I faced the extremely difficult task of penetrating and understanding the fog of claims and counterclaims made by strong advocates on both sides of the issue. All proclaimed lofty ends. No one proposes to demonstrate the evils which will ensue if we, and others, follow our particular right way. No salesman who hopes to sell in volume deliberately sets out to show how use of his product will be detrimental. Those who work with pesticides are salesman for their technological utility.

None of us deny the efficacy or utility of many pesticides. These are well defined and accepted, even by most critics. However it does little good to point out only advantages. We all acknowledge them and continually remind ourselves of them, thereby strengthening our convictions. Such mental exercises become superficial and may ignore the basic value questions raised by use of pesticides which may impair human health. Value questions are being raised by many responsible people who do not a *priori* question the need for pesticides as tools of modern agriculture, but who are concerned about their use and long-term effects. It is common to hear that available evidence indicates that present levels of pesticide residues in our food and environment do not produce adverse effects on the environment or our health. The keywords are 'available evidence.' An issue of concern in the late 1960s and early 1970s was use of the herbicide 2,4,5-T. It was used for broadleaf weed and brush control extensively in Vietnam.

Studies initiated by the National Cancer Institute in 1964 pointed to 2,4,5-T as a possible teratogen. This was new knowledge concerned with the health and welfare of man. Without exploring the many ramifications of the issue and without judging 2,4,5-T, I present some of the evidence. The Dow Chemical Company found no increase in the incidence of birth defects in rats fed 2,4,5-T, containing less than 100 ppm of 1,3,7,8-tetrachlorodibenzoparadioxin (TCDD), a known teratogen, at 1.0–24 mg/kg of body weight/day. The National Institute of Environmental Health Sciences, in a series of experiments with mice and 2,4,5-T, containing 1.0 or 0.5 ppm dioxin, and fed at 50–150 mg/kg of body weight/day, found increasing birth defects and fetal toxicity in mice from 2,4,5-T in the purest form available. Renal defects and excess fetal mortality were observed in rats. The US Food and Drug Administration using 2,4,5-T at 40–100 mg/kg of body weight with several levels of TCDD found increased embryo toxicity and gastric or intestinal hemorrhages in pregnant hamsters. Birth defects consisted chiefly of poor head fusion and absence of eyelids, but were few in number. Analysis of the data show that if one chooses the proper test animal (mice) and a high dosage administered by gavage, teratogenic effects can be shown. The relationship between these data and the likelihood of teratogenicity in the real world was not clear, and it is on this point that the value question about 2,4,5-T hinged. The specter that mandates serious consideration and lends validity to the value question was that cast by thalidomide. The lowest effective human teratogenic dose is 0.5 mg/kg body weight/day. Corresponding values for the mouse, rat, dog, and hamster are 30, 50, 100, and 350 mg/kg body weight/day, respectively. Thus, humans are 60 times more sensitive than mice, 100 times more sensitive than rats, 200 times more sensitive than dogs, and 700 times more sensitive than hamsters.

The Response

The scientific community, regulatory agencies, the chemical industry, and the public have raised their standards concerning safety of pesticides and other synthetic chemicals with known environmental exposure. Legislative machinery was unavailable for definitive action on teratogens, but action was taken and the questions have continued to arise. Dr. L. Dubridge (former science advisor to the President) commended the evolution of thinking about health, but raised the question of where the end point may be. It is the nature of science that as more research is performed, more questions arise. He proposed that decisions will always have to be made on incomplete information and he hoped the judgmental structure and the judged would be sufficiently sophisticated and flexible to accept the knowledge available and permit changes when new knowledge accrues. The Mrak Commission commented that "the provision of food and fiber and good health must be weighed by each country against potential or even actual hazards to health from pesticides. Protection of human health involves a system of priorities which are necessarily different from place to place." Those who work with pesticides base many arguments for their continued use on the premise that pesticides are necessary for production of food and fiber. In general this is a valid assumption, but like most generalizations it is weakened or fails completely in specific situations. The argument often ranks the value of technology in food production ahead of the value of human health.

Under what conditions, would we say that there is any pesticide so necessary that any risk of one of these effects is acceptable? My family would have to be in more than highly theoretical danger of starving before I would allow any risk. We cannot dismiss the question as irrelevant. Those who opposed continued use of 2,4,5-T, based their concern on the data presented. Other examples could be included, but the value question is the same. We can and should publicly deal with such value questions. They should be kept foremost in our minds as new compounds are developed, because there is a great difference between carcinogens and cancer, terata or mutagenic agents and simple poisons. If such changes are induced, they may be irreversible. That is to say no recovery is possible even when the offending agent is removed.

We must continually address ourselves to the real world significance of the valid questions raised. We must yield when wrong but persevere when right. We should all remember to use our intellect and reason. Our intellect should help us distinguish between the possible and impossible; while reasoning should help us distinguish between the sensible and senseless. The possible may be senseless.

Opposition to Pesticides

The general public and an increasing number of the agricultural community oppose pesticides. In light of their undeniable efficacy and the wealth of data supporting their contribution to food production in the world's developed countries, why should this be so? Are these people just opposed to modern agriculture? Do they yearn for a return to "Little House on the Prairie" agriculture without recognizing the drudgery and poverty that went with that kind of agriculture? I do not think so.

Critiques of modern agricultural practice are often superficial, one sided, and lacking in any obvious attempt to understand its complexity. However, it is that very complexity that fosters the critique. Some perceive that modern agriculture has become so complex that it tends to establish its own conditions, to create its own environment and draw us, unknowingly, into it. Agriculture performs an essential function. It produces food and we cannot survive without our daily bread. But we do not live by bread alone. We all need a reason or several reasons for living. Humans want more than just biological existence once our needs for food, clothing, and shelter are met. We want a good life which enables us to realize our potential for human development. We want purpose, hope, and a way to make sense of the meaning of things. Modern agriculture provides abundant food, but works against these other things because man is not included in its design. Agriculture has moved from a position of terrifying ignorance and dependence to a place of knowledge and power (Morison 1966). We are the master manipulators of the environment, which humans, in contrast to all other species, manage to meet our needs. We live where we want and eat what we want to when we want it. Pesticides are part of the means used to achieve our independence from nature and the desirable end of high food production. The modern agricultural system is beyond the comprehension of most citizens. They feel removed from the source of their food which has become just another commodity. Do farmers grow food or just another commodity to sell? Many agriculturalists will say it is the latter and, if that is so, do people eat food or just another commodity they buy? Our techno-logical triumphs in agriculture have produced an environment and a food supply we no longer trust.

People are not so much opposed to pesticides as they are alienated from a system they do not understand and they fear is working against their best interests. Agriculture may need to move away from maximizing production and profits toward maximizing quality and participation. People may know that pesticides are part of modern agriculture, but consumer exposure to them is perceived to be involuntary and people inevitably fear what they do not understand or what they feel is being forced upon them without their participation. We must discover the means to allow society to maintain control of agriculture's nature and direction. Morison (1966) suggests three things may be necessary to achieve this:

1. Members of the society must feel they are participating in the way things are ordered, that they have the power of choice.
2. To make this sense of choosing and participation real, members of society must have available the kind of evidence required to judge possible alternatives.
3. Beyond the evidence supplied for any particular case, people must have a sense of a more general purpose that would serve as a governing context into which particular judgments might be fitted.

These things do not represent the operative paradigm in agriculture, especially in the pesticide realm. People are not fearful of agriculture which has provided such an abundant food supply for so long. They fear that part of agriculture from which they feel alienated. When agricultural practitioners appear to be maximizing production or profit rather than human safety, environmental quality, food quality, soil conservation, or maintenance of rural communities and values, people get uneasy about agriculture and its practitioners. It is not theirs anymore. It is con-trolled by 'them' and they are not 'us.' The mysterious and omnipresent 'they' appear to espouse different values and it does not seem right even though most people cannot articulate exactly what it is that is wrong. Our operative paradigm must be examined to determine why we practice agriculture the way we do. Examination of basic hypotheses may help us understand what happened in agriculture to create unease among the general public.

References

Keen S (1969) To a dancing God. Introduction. Harper and Row, NY, p 1
Morison EE (1966) Men, machines, and modern times. The MIT Press, Cambridge, 235 pp
Zimdahl RL (1972) Pesticides—a value question. Bull Entomol Soc Am June:109–110

Chapter 3
The Pesticide Paradigm

The use of natural and synthetic chemicals as pesticides is an ancient agricultural practice. In 1000 B.C., Homer wrote of the pest averting sulphur. In 470 B.C., Democritus suggested that residues from the production of olive oil could be used to cure blight. The harmful effects of salt were mentioned by Xenophon in 400 B.C. and the Romans sowed their enemies' fields with salt as continuing punishment (Smith and Secoy 1976). Mercurous chloride was first used as a fungicide for seed treatment in 1755 and Bordeaux mixture (copper sulphate, lime and water) was discovered in France in 1865. It was used to control downy mildew on grapevines. Selective control of weeds began around 1900 in France, Germany and the US using sulphates and nitrates of heavy metals. The first synthetic organic chemicals were introduced in 1932 (2-methyl-4,6-dinitrophenol for weed control) and in 1934 the first patent on dithiocarbamates as fungicides was granted.

None of these gained the widespread use and commercial development of DDT during WW II and 2,4-D afterward. They were inexpensive to produce, easy to use, provided excellent pest control and were apparently safe for all. Their development and use and the hundreds that have followed have led to the currently operative pesticide paradigm discussed in this paper.

Paradigm

Paradigm comes from the Greek word "Paradeigma" meaning a pattern or example. Kuhn (1970) defined a scientific paradigm as "A universally recognized scientific achievement that for a time provides model problems and solutions to a

Adapted with permission from the Bulletin of the Entomological Society of America. Zimdahl, R. L. 24:357–360. 1978.

community of practitioners."For Kuhn, the paradigm is not a set of answers or a description of the ultimate destination. It is similar to a road sign, which clearly indicates direction, but provides options about the route. The paradigm does not replace the scientific method nor explain all questions; it defines appropriate questions. For the scientist, the paradigm and the questions it poses are operative for a time, always subject to discussion, further articulation or drastic alteration. A divergence of opinion and methods is to be expected. All members of a scientific community work within the bounds of a paradigm which guides their research even though it may not be fully interpreted. Such interpretation often awaits perception of inadequacy or incompleteness.

The community of pesticide scientists is no exception to this generalization. It is an identifiable group because they share common terminology, read the same or similar journals, attend the same kinds of meetings and have received similar training. Their paradigm rests firmly on the historical use of many different pesticides from sulphur to synthetic organic chemicals and the numerous research possibilities their advent introduced. Kuhn states, and his book supports, the contention that "the road to a firm research consensus is extraordinarily arduous." I submit that pesticide scientists, often trained in other disciplines, have achieved a research consensus, and thus a paradigm, which should be explored.

Pest

Many dictionaries define a pest as a plant or animal detrimental to man or his interests. The human element of the definition has been particularly important in the evolution of the pesticide paradigm. That is, pests have been defined by man, not by nature. Nature knows no such category. There is vigorous competition for survival in nature. There are species which succeed and those that fail, but only humans consciously attempt and succeed in manipulating the environment to their advantage. All other species must adapt to the environment or die.

It is man who has created both the category of pest and the tools, from bulldozers to pesticides and fly swatters, which permit manipulation of the environment. We have assumed a right, which some would term an obligation to control. While this concept is commonly accepted by pesticide scientists, it is not commonly recalled during the practice of pest control. Although the general definition of pest has not changed, the rising level of concern for maintenance of environmental quality has raised probing questions about its acceptability. It is no longer sufficient to identify a pest. One must carefully define when and where it is a pest and be cognizant that a reduction of pest populations to increase crop yield or protect human health may be the desired end, but other ends are also probable. The definition cannot stop with the general definition of pest which implies a right to control. We must realize the ability to control does not confer a right to control. We need to explore all ramifications of control as well as the tools of control.

The Pesticide Paradigm

The pesticide paradigm includes two fundamental propositions. The first states that there are species that should be classified as pests and that it is necessary to control their populations to produce food and maintain human health and comfort. This part of the paradigm is widely shared because most people really do not like crabgrass (weed pest) or soft rot of potatoes (bacterial pest). People are also irritated by mosquitoes (insect pest) and their health can be irreparably damaged if those mosquitoes carry malaria protozoans (bacterial pest) or yellow fever (viral pest).

The second proposition is the less widely accepted inclusion of pesticides as primary weapons in the arsenal of pest control technology. Many would say that pest control exhibits a dependence on pesticides. This accusation usually devolves to an argument rather than a discussion and becomes an irreconcilable issue between proponents of opposite views. It is not my purpose to defend either of these views, but rather to examine the basis of what I have posited as the two components of the pesticide paradigm and to examine how the paradigm may change.

The advent of synthetic organic pesticides after World War II permitted the development of action programs to control or attempt to eradicate pest populations. Because of the rapid development of pesticides the perceived need to control could be achieved. The species man defined as pests became objects to be studied, understood (known), and controlled. Studies to understand pest species were often subordinated to studies of pest control. Because of the ability to control, pests could no longer be regarded as components of the environment but as raw material. These studies were good, productive, and necessary science but their objectives were made possible by the development of pesticides.

Knowing the Issue

Pest control studies fit within Kuhn's (1970) definition of normal science, i.e., "the activity in which most scientists spend almost all their time." Normal science is "predicated on the assumption that the scientific community knows what the world is like." Tillich (1951) has dealt with cognitive relations, what it means to know something, and points out that knowing is an act of union of the subject (the knower) and object (that to be known). Subjects observe objects, interpret them and fit them into their frame of reference. However, the union is an anomaly in that it requires separation or detachment. In order to know one must look at an object and looking requires separation, that Tillich terms a 'cognitive distance.' Tillich calls the knowledge resulting from separation 'controlling knowledge' which he contrasts with 'receiving knowledge': that which emphasizes creation of union between subject and object. The definition of pest, which separates, while our ability to control has made pests objects of controlling knowledge. While mankind has strenuously resisted becoming objectified, we have made pests completely conditioned and calculable things deprived of their subjective qualities. They are objects of the knowledge of technical reason which looks upon them and, in a metaphorical sense, does not see them looking back. Our relationship to pests is one of dominion. It is our destiny to subdue (control) them so the environment will conform to our design. If there is wisdom in the ecology of nature, one must question the possible ends when pests are objectified as they are when perceived only through controlling knowledge.

In contrast, receiving knowledge includes an emotional element that controlling knowledge tries to avoid. Emotion, or the expression of emotional preferences (i.e., one kind of values), is the vehicle for obtaining knowledge in the creative union but Tillich notes, the vehicle does not make the content emotional. Content may be rational, verifiable, and received with critical action. However, emotion is a prerequisite to receiving knowledge because union of subject and object is impossible without emotional participation. It is through union that we achieve logical meaning or understanding, which "involves an amalgamation of controlling and receiving knowledge, of union and detachment, of participation and analysis." (Tillich 1951)

Controlling knowledge dominates the pesticide paradigm and the practice of pest control. Controlling knowledge is precise, publicly verifiable and undoubtedly successful in terms of pest control. Great stores of empirical knowledge have been produced, and research continues to develop knowledge about pests and how to control them. However, controlling knowledge, which only objectifies, runs the risk of losing sight of the element of union and perhaps of real understanding. Pests become what we, and our controlling knowledge, consider them to be. They exist, therefore, they must be controlled. But do they exist only to be controlled? To know how is not necessarily to know why! What is the natural role of what man calls pests, and where is the emotional element in knowing pests and what to do about them? Pesticide science, working within the proposed paradigm, presents the truth about pests as that contained in empirically verifiable statements, but truth is not restricted to experimental verification and testing by repetition. "Truth is the essence of things as well as the cognitive act in which their essence is grasped."(Tillich) The pursuit of truth requires controlling and receiving knowledge.

One test of truth is experimental, which the pesticide paradigm has tended to embrace as the only method. It has avoided or ignored experiential verification as a test of the validity of pest control or pesticide use. Experiential verification, derived from cognitive union, has usually been labeled as environmentalism or emotional (which it is) as opposed to scientific and factual (which it also may be). Such verification often suggests that pests are not just objects to be controlled and that ends other than control are the result of pesticide use. The technical success of pesticides and pest control is an impressive verification of the success of controlling knowledge of pests. Receiving knowledge is not as easily or as precisely verifiable. Life makes the test of verification and it takes longer, is not controllable, and has an element of risk. Tillich (1951) said "life processes have the character of totality, spontaneity and individuality." Whereas "experiments presuppose isolation, regularity and generality." Therefore, only separable elements of life are open to experimental verification while life itself must be received in creative union and verified experientially to be understood.

Thus, within the pesticide paradigm, pests are objects to be controlled. They are Objects of controlling knowledge, and pesticides are a means of control. The rise of environmental concern, awareness of the interdependence of life and a great regard for all creatures great and small[1] has brought the pesticide paradigm into question. We should recognize that controlling knowledge, which dominated the paradigm, is verifiable but may not be ultimately significant, while receiving knowledge can be ultimately significant but cannot give the same degree of certainty.

The Paradigm Shift

The shift that is occurring is not tantamount to a revolution. Pests are still recognized as such and the need for their control will continue. The most important questions concern the application of the definition of pest and the use of pesticides as the primary means of control. Lowrance (1976) accurately describes the questions and discusses the limitations of science in determining answers. His book begins by stating some assumptions which most pesticide scientists would accept:

- Technology has not been an unmixed blessing and will remain with us.
- Many of our problems are technological in origin but will necessarily be solved in the political, and not just the technological realm.
- Human activity will always and unavoidably involve risk.
- To make a safer world we can start changing only from where we are now.

The development of pesticide technology has brought undesirable side effects as well as great benefits. In spite of conflicting opinion about the relative magnitude of the benefits and risks[2] pesticide use and development will continue for the foreseeable future and only the unknowledgeable would suggest it stop entirely. Such use will occur in a world of changing values and expectations, a world where know-why will assume equal or greater importance than know-how. What one person regards as a weed, an object of controlling knowledge, another may see as a wild flower, an object of receiving knowledge; to be appreciated,

---

[1] The title of a best seller. Herriot, J. (1972) All Creatures Great and Small. Bantam Books, 499 pp. Title derived from the well-known hymn All Things Bright and Beautiful by Cecil Alexander.

[2] See Atreya et al. (2011).

enjoyed, and lived with. While one person perceives insects as pests others will see a natural order with which we must not tamper casually.

We all want a good environment. But who has the right or knowledge to define what is good? The history of pesticide legislation illustrates that those who develop pesticides, those who use them, and the scientists who think they know pests and their control best, are not the sole possessors of the rules of the game. The public and the political process are involved and probably will become more so. The definition of what a pest is and when it is a pest is not solely the province of the pesticide developer, the scientist, or the pestered. Nor do the pestered have the right to invoke a control of their own choosing. The use of pesticides is subject to public review and this creates pressure to examine and possibly change the paradigm.

During the decades of pesticide development, public health concern in the US has changed from microbiological to micro-chemical. All who are involved with pesticides are aware of probing questions about carcinogenicity, mutagenicity, and teratogenicity. The US Environmental Protection Agency has promulgated controversial rules for evaluation of the carcinogenicity of pesticides in development. During the ensuing discussion about environmental safety of pesticides, it has become clear that many miss a point that Lowrance (1976) has made clear. The business of the scientist is to answer scientific questions and seek scientific truth. The business of the advocate is to win without violating the law. The pesticide environmental issue has often become an adversarial procedure in which some pesticide scientists bemoan the disregard of what they perceive as facts when the issues seem to be consistently decided by advocates of complete human or environmental safety with little regard for the probability of untoward events and apparent neglect of pesticide benefits. What these scientists fail to see, as Lowrance clearly illustrates, is that safety cannot be measured, only risk can be measured. Something is safe if its attendant risks are judged to be acceptable. Measuring risk is an empirical scientific activity, but judging safety is a normative political activity. Pesticide risks are measured scientifically, but judgment of safety has been, and should be, made in full consideration of all available ethical, economic, political, and scientific information. The science, i.e., the measurement of risk, may become a secondary or lower issue when the political realm judges safety and thus acceptability. Such judgments will be made and many current debates center on acute vs. chronic effects. The use of pesticides in the environment has always been predicated on two assumptions (Alexander 1965):

1. People must eat (and remain healthy).
2. People should not be poisoned now or in the future because they cannot avoid eating.

It has been relatively easy to answer acute toxicity questions. Obviously pesticides (from pesta L—pest, and caedere L—to kill) are poisons. If they were not poisonous to something they would be useless. There are threshold dosages for pesticides and mild and serious effects are different for each. Acute toxicity levels can be determined in laboratory studies and during use, poisoning can be avoided.

However, the chronic toxicity question is more complex. Scientists can make predictions, but cannot offer proof that the present use of a pesticide will not cause untoward effects at some unknown future date. Predictions of the course of events in the remote future are a prophetic not a scientific activity.

Pesticide manufacturers are required by law to test the carcinogenic and teratogenic properties of products prior to public sale. These tests are conducted on laboratory animals at high dosages. In these tests, is a compound that produces no tumors of any kind in 100 animals or one that produces only benign tumors to be considered non-carcinogenic? Scientists do not agree on the answer.[3] However, the US Environmental Protection Agency (EPA) has decided that any risk is too great whether it produces benign or malignant tumors. EPA has promulgated and is presumably acting on the basis of its principles of carcinogenicity (Ling-Yee 1975) which, among other things, state that: A carcinogen is any agent that increases benign or malignant tumor induction in man or animals. The concept of a threshold exposure level for a carcinogenic agent has no practical significance because there is no valid method of establishing such a level. Editorials and a series of letters in Chemical and Engineering News (1975) have argued the issue and made it abundantly clear that it is unresolved.

These are normative political judgments of safety that have considered the available science, but are not scientific measurements of risk. Such judgments are made, usually with incomplete data[4] in what is understood as benefit risk analysis. This frequently discussed method often includes qualitative judgments as well as quantitative analysis on questions of the relationship of pesticides to environmental or human safety. An aspect of the problem is that judging unpredictable risks of chronic toxicity against benefits which may be equivocal is often a matter of relying on the opinion of one group of experts as opposed to another, equally well trained and competent group. EPA traditionally has sided with safety, much to the consternation of many who view this as overemphasis on risk. They see pesticides as guilty until proven innocent and this is consistent with the view that the major pressures retarding economic development are those on the side of safety (Kahn et al. 1961).

Kahn et al. point out that people are more willing to accept deaths (and by implication environmental damage), which are not traceable to specific causes, but only when they cannot identify victims ahead of time, and prevent death. Political decision makers and most scientists have come to the conclusion that damage over time is just as bad as damage in a short time and find this unacceptable. Thus, we find ourselves working within a paradigm that generates controversy between

---

[3] If no tumors are found in 100 animals, statistical analysis shows that at a 95% confidence level the tumor incidence is less than 4% not O. [See Plant A. F. letters to the Editor. C & E News (1975)].

[4] This comment should not be interpreted as a criticism of regulatory agencies or of the scientific method. Action, decisions are commonly urged by interested parties but science, with its incomplete understanding of the functioning of most biological systems, leaves many questions unanswered.

unpredictable risks (chronic toxicity) and benefits which have not been elaborated to the satisfaction of the larger society. Any modern society subjects its citizens to all manner of risks some of which are great. Citizens who want the benefits of that society accept the risks as part of living. We drive cars, a dangerous activity, but it is not regarded as very dangerous because we are in control and we (at least I) are good, careful drivers. We also fly in airplanes, which is usually regarded as a more dangerous activity. However, the per hour death rate for driving vs flying is about equal.[5] But for most people there is something particularly insidious, especially dangerous about the unknown risk of extremely small amounts of pesticides that might be in our food, air, or water. It is generally accepted that, for most people, they cause few acute effects. Still people wonder if they will get cancer in 20 years? This is an obviously unanswerable, but nagging question. The majority of society seems to support what EPA does about regulating pesticides, especially when it concerns unknown but potentially dangerous chronic effects.

Kahn et al. (1961) do not question the wisdom of siding with safety. They provide an example which those who judge should consider. They ask if damage over time is as serious as damage that occurs in one generation. Their answer using the following example suggests it is not. "Imagine that society must choose between four situations:

1. 100% of the next generation would be killed.
2. 0% of the next 10 generations would be killed.
3. 1% of the next 100 generations would be killed.
4. 0.1% of the next 1,000 generations would be killed.

The first case is the end of history, everybody dies. In the second case there is no problem and thus, no concern. In the last, and perhaps the third, case, great damage occurs, yet it is scarcely apparent because it is spread out over such a long time and among so many people. Clearly the first choice is intolerable. The fourth, while tragic and nasty, would certainly be tolerated better under most circumstances. Indeed in many situations similar to the fourth case, it would not be possible to measure the damage or prove that it existed. Any analysis of the difference between the first and fourth situation must take account of this spread over time, even though the total number of people killed is exactly the same."

Benefit-risk questions concerning pesticides will be with us for some time, it will behoove us to recognize them and develop answers appropriate to the questions evolving from a new paradigm. As we do, we should be cautioned by the words of Laski (1930):

> Special knowledge and the highly trained mind produce their own limitations. Expertise, it may be argued, sacrifices the insight of common sense to intensity of experience. It breeds an inability to accept new views from the very depth of its preoccupation with its own

---

[5]   It is true that a few more than 37,000 people were killed on US highways in 2009 and far fewer died in airplane accidents. We spend more time in cars than in airplanes and when that difference is considered the per hour death rate is about equal (Levitt and Dubner 2009, p. 151).

conclusions. It often fails to see round its subject. It sees results out of perspective by making them the center of relevance to which all other results must be related. Too often, also, it lacks humility; and this breeds in its possessors a failure in proportion which makes them fail to see the obvious which is before their very noses. It has, also, a certain caste spirit about it, so that experts tend to neglect all evidence which does not come from those who belong to their own ranks. Above all, perhaps, and this most urgently where human problems are concerned, the expert fails to see that every judgment he makes not purely factual in nature brings with it a scheme of values which has no special validity about it. He tends to confuse the importance of his facts with the importance of what he proposes to do about them.

How Will the Paradigm Evolve?

There are some clear signals that indicate a direction of evolution if not an end point. If one accepts that agriculture will be practiced on the soil in the open environment for some time to come, there is no reasonable justification for claiming that pests will not continue to hinder food production. If pests remain, some means for their control must remain. The present method of choice in many, but not all, situations is a pesticide and it will continue to be so for the foreseeable future. If one envisions agriculture evolving away from the open environment or away from soil[6] the evolution of the pesticide paradigm could occur in a very different way. I suggest the first, presently more plausible, assumption will be dominant for several decades.

Signals indicate there will be more governmental regulation and greater emphasis on safety and proof of benefits. This may lead manufacturers to develop pesticides that have more specific action against a single pest, growth stage, or biochemical process. New application methods will also be developed to permit precise, accurate placement of pesticides on targets and strict avoidance of spray drift and contamination of non-target organisms. A second development furthered by increasing regulation and knowledge is the effort by manufacturers to develop specific biochemical inhibitors. Future pesticides will, in all likelihood, be developed by first knowing what biochemical reaction can be blocked to cause pest death or malfunction and then designing a pesticide to do that. Further development would center on formulation and application of the known biochemical inhibitor. Currently, the empirical screening method dominates; many candidate chemicals are tested to see if one is active and selective against a pest or series of pests. Only when this is determined, does one ask how it works. Empirical

---

[6] Although irrational to many this is not, strictly speaking, impossible. One must consider the pressure of rapidly expanding populations on agricultural land and other relevant events such as cheap desalinization of seawater and what that might do to develop hydroponic agriculture (Kahn et al. 1961). Vertical farms in city centers where plants are grown hydroponically have been proposed (Anonymous 2010). Questions of production efficiency and profitability have not yet been answered satisfactorily.

screening is a good method. It has been the best available. Its replacement does not depend on the willingness of manufacturers to adopt the method proposed, but rather on the development of sufficient biochemical data so new methods can be used. The reason we do not have an herbicide that interferes with cellulose synthesis (a specific plant process) may be that we do not know precisely how plants synthesize cellulose.

The paradigm also will be modified by greater use of what pesticide scientists know as integrated control. This is defined as the use of cultural, mechanical, chemical, and biological control methods, in an appropriate combination, to maximize pest control. Integrated pest/weed management is a very good idea. It has been difficult to implement because it is not known how each practice or their many combinations will affect specific pest populations in all environments. As this knowledge is developed most pesticide scientists are confident that integrated control will reduce (but may not eliminate) pesticide use. Finally, I think pests must be increasingly understood, not as objects of controlling knowledge, but as objects of receiving knowledge; living things with which we share an environment. The paradigm of pesticide science may be evolving in the same direction as medical science. Some pests (e.g., smallpox) are regarded as so serious that eradication is the only acceptable answer. Other pests (e.g., poliomyelitis virus) are regarded as so serious that they must be prevented and eradicated, when possible. The physician's paradigm, and knowledge, permit practice of preventive medicine rather than just curative therapy.

Pesticide scientists have been constrained by limited knowledge of pest and host biochemical function. They are also constrained by a paradigm that defines pests as objects to be controlled and subjugates understanding to control. This forces the practice of curative rather than preventive therapy. As knowledge progresses and the limitations of our value system are recognized, the pesticide paradigm should evolve in the direction suggested by Naegele (1993). He hoped that Weed Science would "overcome the paralysis of the pesticide paradigm and conceive a weed science research program that addresses both society's perception of safety and the scientific community's perception of risk."

A justification for evolution of the pesticide paradigm is that—we are not independent living things, surrounded by other living things that are here for our benefit. We are part of a living system over which we have not been given dominion. We are charged to care. We cannot continue to harm ecological and social communities and survive on the only planet we have. The necessity of thought about the paradigm was expressed well by Martin Luther King, Jr. (1967) in a speech not long before he was assassinated.

> We are now faced with the fact, my friends, that tomorrow is today. We are confronted with the fierce urgency of now. In this unfolding conundrum of life and history, there is such a thing as being too late. Procrastination is still the thief of time. Life often leaves us standing bare, naked, and dejected with a lost opportunity. The tide in the affairs of men does not remain at flood–it ebbs. We may cry out desperately for time to pause in her passage, but time is adamant to every plea and rushes on. Over the bleached bones and jumbled residues of numerous civilizations are written the pathetic words, "Too late."

References

Alexander M (1965) Persistence and biological reactions of pesticides in soils. Soil Sci Soc Am Proc 29, 1

Anonymous (2010) Vertical farming-does it really stack up? Econ Technol Q Dec 11:15–16

Atreya, KBK Sitauala, Johnson FH, and Bajracharya RM (2011) Continuing issues in the limitations of pesticide use in developing countries. J Agric Environ Ethics 24:49–62

Chem. Eng. News. Plant, AF Sept. 22 and Oct. 20, (1975) Editorials, and Letters to the Editor. Byer, AJ Dec. 15, '75; Stemmie, JT Dec. 22, '75; Weinhouse, S Jan 26, '76; Blair, EH Feb 9, '76; Pike, EA and Walker, W Feb 23, '76 (2ltrs); Wagner, BM, Brame Jr, EG, Dole, GF and Wald,G Mar. 15, '76 (4 ltrs); Scribner, JD Apr. 19, '76; Kolin, P May 10, '76; Vallaire, C June 14, '76, Weinhouse, S July 5, '76

Kahn H, Brown W, Martel L (1961) The next 200 years: a scenario for america and the world. W. Morrow and Co. Inc., New York, p 241

King ML Jr (1967) Beyond vietnam: a time to break silence: declaration of independence from the war in vietnam. April 1967, Manhattan's Riverside Church

Kuhn TS (1970) The structure of scientific revolutions, 2nd edn. International Encyclopedia of Unified Science, vol. II, No. 2, 210 pp

Laski H (1930) Diderot: Homage to a genius. Harper's Mag 162:597–606

Levitt SD, Dubner SJ (2009) Freakonomics. a rouge economist explores the hidden side of everything. Harper Perennial, New York, 315 pp

Ling-Yee G (1975) Chem Eng News. Nov. 3, 1975

Lowrance WW (1976) Of acceptable risk. Science and the determination of safety. W. Kaufmann, Inc., Los Altos, p 180

Smith AE, Secoy DM (1976) Salt as a pesticide, manure, and seed steep. Agric History 50:506–516

Tillich P (1951) Systematic theology, vol I. University of Chicago Press, Chicago, 300 pp

Zimdahl RL (1978) The pesticide paradigm Bull Entomol Soc Am 24:357–360

References

Chapter 4
A Question of Faith

The original purpose of this essay was:

1. To describe and discuss the history of weed science (Chap. 1).
2. To discuss value questions related to pesticides (Chap. 2).
3. To present my view of weed science's dominant paradigm (Chap. 3).
4. To question the faith of many involved in agriculture in the possibility of perpetual increases in production and ever more efficient herbicides. This edition has been edited using the perspective gained over 20 years but each chapter retains the same purpose.

Several writers, usually regarded as pessimists, have questioned the Western world's unalloyed faith in technology and the possibility of continual increases in food production (Berry 1981; Black 1970, Brown 2004; Durning 1996; Ehrlich and Ehrlich 1996; Jackson 1980; McKibben 2003, and The WorldWatch Institute). Other, equally competent, but far more optimistic writers regard technological progress as inevitable and the only sure route to feeding the growing world population (Bailey 1995; Capra 1982; Gore 2006; Lomborg 1998; Ridley 2010; Simon 1995, 1981, Simon and Khan 1984; Smil 2000, and White 1968). These authors share the view that problems (e.g., feeding a growing population, pollution, loss of species and agricultural land, poverty, diminishing water and mineral resources) that dominate the thinking of pessimists can be solved. Human ingenuity is the required resource. Both groups regard themselves as neither optimists nor pessimists, but as rational thinkers. Since 1800 the indisputable evidence shows that the majority of the earth's population is better fed, better sheltered, protected from disease, richer and lives longer. The optimistic view prevails and affects how agriculture is practiced in developed and developing nations.

This chapter uses weed science and herbicides, a major part of western agricultural technology, as a window through which one can view the future of agriculture and its technology. My hope, my plea is that as the future comes into view, it will encourage thought about whether the optimists or pessimists

R. L. Zimdahl, *Weed Science: A Plea for Thought*—Revisited,
SpringerBriefs in Agriculture, DOI: 10.1007/978-94-007-2088-6_4,
© Robert L. Zimdahl 2012

are correct. A careful view should help all think about the future of humanity and the conventional agricultural wisdom about the technological route to greater production. The conventional wisdom about anything is that which:

> sets great store by what it calls constructive criticism. And it reserves its scorn for what it is likely to term a purely destructive or negative position. In this, as so often, it manifests a sound instinct for self-preservation. The attack on the conventional or accepted thought is dismissed as an inferior and, indeed, a wanton activity and, as such not something that should be taken seriously.

> Galbraith 1958

Within agriculture and weed science, the conventional wisdom is that herbicides and all pesticides are useful, necessary, tools of modern technology and have been and can be used intelligently. It is often pointed out that the problem is not the pesticide per se, it is the people who use them. The argument that guns don't kill people, people do, is similar. These, often unexamined, points of view are framed and bound by experience and reinforced by the supporting community. Those secure in their possession of the conventional wisdom are not inclined to solicit or accept contrary views. It is frequently a biased examination of selected scientific data or opinions about what the data say without recognizing that science is not capable of answering all questions and that data are developed by practitioners who are not always aware of their own paradigm and its inevitable bias.[1]

This essay is an argument that addresses the conventional wisdom about pesticides. It is not an attack on data about pesticides. It is a discussion of attitudes toward data and conclusions drawn. Indeed, part of the basic premise of this essay is that arguments are not just about data. Arguments are often about subjective interpretations of data and these should not always be dismissed because they lack objectivity. Subjectivity appears on both sides of the issue. The essay claims that the basis of many arguments is a difference in attitudes and thinking, what can be called the mind set of agriculture and agriculturalists.

Pesticide is a word which evokes many images, especially concerning their associated risks. College students ranked them fourth on a list of 30 most risky items in 1982. Members of the League of Women Voters ranked them ninth, and business and professional club members placed them fifteenth. Actually, when the number of accidents per year was considered, pesticides ranked 28th out of 30 known risks; between food preservatives and prescriptions.

The number of accidents recorded per year is a precise, quantitative measure and those who know how can evaluate hazards and measure risk based on such quantitative data. Most people, however, rely on intuitive risk assessment which, by definition, is not a precise quantitative measure. Their information comes from news media, friends, and most certainly, just feelings (Slovic 1987). The dominant feeling among Americans is that they are more at risk from pesticides today than

---

[1] A bias is a preference, uninformed or unintentional inclination, especially one that inhibits impartial judgment. Those with a bias are often perceived as being against something. However, a bias may be for or against something.

they have ever been before. Intutively, caring and fear are focused on ourselves and those we love. Pesticides are regarded as a greater risk to ourselves than other environmental hazards, although risk to others is acknowledged. In 1991, people felt and most still feel, that the risk to humans is high, which is in marked contrast to how professional risk assessors view pesticides (Slovic). They claim that manufacturers and users pay careful attention to risk reduction whereas the public sees the exact opposite. Defenders of pesticides observe the enormous effort devoted to assuring that nothing goes wrong and to making sure things go right.

Part of the problem is that people don't believe the numbers. They don't believe that risk assessors and obvious advocates present objective evidence. Tragedies, such as airplane crashes, fish kills from pesticides, or the threat of AIDS overwhelm most people. Actual data aren't as influential as real and truly scary events. These lead to assumptions that life is getting riskier. Also people find it difficult to acquire and comprehend data on pesticide safety. Such data are not published in the things Mr. and Ms. citizen routinely read or see on television. The data, based on tests on laboratory animals and important assumptions about unity of biochemical response among species, involve extrapolation of results from small test populations of animals to large human populations. When presented with such data people don't understand them nor do they easily accept proposed actions based on them.

The public is further confused and made suspicious when they learn scientists don't agree on the validity of underlying assumptions and have contrary interpretations of the data. The public's competence and wisdom are questioned by rational scientists who know and believe the underlying assumptions and support proposed actions. Scientists see irrational fear, not supported by the data, from people who smoke, overeat, and consume too much alcohol. The scientifically rational mind wonders why those who ignore obvious health effects of known hazards are so fearful of potentially less harmful but not completely described effects of hazards like pesticides. They wonder if Mr. and Ms. citizen are really that irrational? Perhaps their attribution of unknown, but possibly large potential damage from pollutants like pesticides alleviates personal responsibility for a poor life-style or relates poor health to unavoidable external pollutants and relieves one of direct responsibility for personal and environmental health. It may also be true that the data are presented in such a way that they are incomprehensible to most people and incomprehensibility increases fear. Finally, there is a fear of involuntary contamination and possible consumption of pesticides with uncertain effects, but acknowledged toxicity. The risks that kill people are very different from the risks that scare them.

The problem is also related to the fact that disagreements about risk do not always disappear in light of new scientific evidence. Strong feelings are not always overwhelmed by contrary scientific evidence. Evidence which agrees with one's preconceived notions is usually immediately accepted and science's credibility is enhanced. But when scientific evidence is contrary to personal, deep seated emotional beliefs it is often dismissed as erroneous, unreliable, or biased by its source of funding (Slovic 1987). Entrenched societal and culturally based beliefs are difficult to change when they affect personal health and safety.

The Quality of Evidence

In 1917, Gustav Holst composed The Planets, a musical description of seven of the nine planets (omitting Earth and Pluto) in our solar system. An image of each planet is created in the mind. The music evokes thoughts, feelings, and emotions. What we hear is common to all listeners but what feelings or images the music evokes varies with each listener's prior perception of the planet and an innate or acquired ability to appreciate music. Not all listeners feel the same way or create the same mental images even when they listen together. Holst portrays Mars as the bringer of war who is forceful and assertive. Venus is the bringer of peace and a lover of all beautiful things. Mercury is musically the winged messenger, quick in thought and ingenuousness. Jupiter is buoyant and hopeful, the bringer of jolliness. Saturn is patient and enduring, the bringer of old age. Uranus is the magician, abrupt, eccentric and unexpected, while Neptune is the subtle, mysterious mystic. Each image was inspired by stories of the planet's astrological character and not by any hard scientific evidence of what they are really like. One is inclined to accept Holst's vision of the planets because most people don't know them well and are willing to let their imagination be led by Holst's music. We revel in our emotions and ignore scientific evidence to the contrary. In the scientific realm, we are prevented, by tradition and the scientific ethos, from giving our emotions or the emotions of others credence on strictly scientific issues. Proponents argue that pesticides are creations of science and must be judged scientifically, not emotionally. This rule is applied to nearly all questions about pesticides and other evidence is dismissed. Weed scientists are, perhaps unknowingly, converts to logical positivism which tests all statements by reference to experience or the structure of language. Critiques which don't fit the language or experiential expectation are rejected. This creates separation from other opinions and dismisses outside, non-scientific, criticism. This separation did not begin with weed science or agriculture. It has been and will continue to be a scientific and societal reality and a problem.

Mayer and Mayer (1974) offered an insightful and still relevant commentary that began with a premise, acceptable to most agriculturalists: most Americans are ignorant of agriculture. I suggest they still are and Americans regard agriculture as a quaint but not wholly necessary enterprise. After all the grocery stores are always full. Agriculture is the most basic human activity and an essential science, but it is isolated, and in large measure has isolated itself, from the mainstream of American life and from intellectual life. Agriculture's success is fundamental to the continuation of our life-style, our economic system, balance of payments, and of most importance, our survival. We must eat to live. American agriculture takes pride in the frequent assertion that in 1990 one American farmer fed himself and over

50 other Americans plus about 30 (depending on whose estimate you want to believe) non-Americans to whom we sell, or occasionally give, our surplus food. In 1900, 29 million farmers fed 76 million Americans, whereas in 2010 approximately 800,000 American farmers fed themselves, more than 300 million other Americans (1 feeds about 150) and through the US agency for international development (US AID) a number of non-Americans. US farmers are among the most productive in the world.[2] Ours is an abundant agriculture, but Mayer and Mayer (1974) state that its success has been built on a remarkably well-integrated, self-contained system that has become a model for a world wanting agricultural development, but may prove to be a tragedy for the United States. The potential tragedy can be attributed to the unsustainable nature of our highly productive agricultural system. Agriculture is the largest and most widespread human interaction with the environment, albeit one with acknowledged undesirable effects. It is not true in the US, but worldwide agriculture employs more people than any other human activity. Much of the work in developing countries, especially weeding, is done by women. Modern chemical, energy, and capital intensive agriculture will continue for the foreseeable future because of its productivity and because its indirect costs (e.g. chemical pollution, soil erosion, loss of genetic diversity, loss of small farms) are externalized. Consequently these costs are borne by the general public and not by manufacturers or users of agricultural technology. The tragedy is that externalization of real costs acts as a sedative that allows continuation of agriculture's March toward environmental catastrophe rather than sustainability. Scientists and farmers are able, if they think about the future of the highly productive system they have developed and benefit from, to avoid the implications of scientific confidence and existing moral dilemmas. The current economic disaster of agricultural overproduction is a triumph of applied agricultural science and a prime example of the simultaneous success and fallibility of technology. United States agricultural science and technology, the land grant system of agricultural research universities and experiment stations, and our agricultural extension system are models for many of the world's countries, but are also a cause of our present over production, the lack of sustainability, and the farm crisis. Science and technology, government policy or lack thereof, and social forces are causes of agriculture's problems. The scientific system that helped create these problems does not accept any blame for it and therein is part of the tragedy. It is an example of the agricultural mind set and justifies Mayer and Mayer's accusation.

Their second claim does not mesh with the experience or goals of those engaged in agriculture and therefore, has not been acknowledged, analyzed and debated. The claim is that the integration and isolation of the system have led to what Mayer and Mayer call-The Island Empire. They suggest that agriculture is an intellectual and institutional island. It is a vast, wealthy, and powerful island, but

---

[2]   Remarks by T. Vilsack, Secretary US Department of Agriculture, March 5, 2010. Anaheim, CA. See: http://www.usda.gov/wps/portal/usda/usdahome?contentidonly=true&contentid=2010/03/0121.xml. Accessed November 30, 2010.

definitely an island. As US agriculture developed in the nineteenth century, it did so independent of other developments and traditions. Agriculture created its own federal department, extension service, research establishment, professional and trade organizations, publications, and public constituency. These have all endured and enforced agriculture's isolation from mainstream American life. The problem now is exacerbated by a declining US farm population and consequent erosion of the formerly strong and cohesive political base associated with agriculture. It is now safe for an aspiring national politician to ignore the ever diminishing agricultural constituency.

Mayer and Mayer strongly accuse American agricultural colleges of being anomalies within their own universities. They are separated from the mainstream of American scientific thought and national discussions about social policy. Because of agriculture's isolation, it does not ask for or receive outside criticism. That situation has contributed to the pesticide paradigm (Zimdahl 1978), unquestioning devotion to pesticides as the appropriate cure for all pest problems. When outside criticism of agricultural practice has been offered, it is rarely regarded, within the agricultural community, as constructive and often its validity is denied because the critic is not an agriculturalist (Mayer is a nutritionist) and, therefore, does not and cannot understand agriculture. Agriculture thus proceeds on unquestioned assumptions that are rigorously challenged by external questioners. This is as true within agricultural colleges as it is among farmers, ranchers, and employees of agribusiness companies. The result is isolation of agriculture's specialists and practitioners from science and the liberal arts the mainstream of intellectual life, within their own self reinforcing organizations. As Berry (1977) and Laski (1930) suggested specialists place themselves in charge of one possibility by leaving out all others, make rigid, exclusive boundaries and reinforce their image of control. Thus, the fiction of absolute control of the agricultural environment becomes possible. However, as Kirschenmann (2010) points out, domination has not given us dominion. Our scientific power and technological wizardry, our Promethean power, have not given us control over nature, which if we are honest, we do not understand in all its complexity. The myth of absolute control is an aspect of agricultural thought that should be examined. Agriculture, and particularly specialists who work with pesticides, must recognize that the fiction of absolute control may inhibit progress of the world's most important human activity—food production. The fiction of control raises two questions which remain unexplored within agriculture. First, one must question if it is possible to control all species but the crop in any field? Twenty years ago the answer was—not usually. More selective herbicides and genetic modification of crops have changed the answer to, yes, at least in the short run.[3] Weed scientists have not asked if they ought to do what is technologically possible? Is it right for

---

[3] A good, but not the only, example of how this goal has been achieved is Monsanto's development of several Roundup Ready™ crops. Until resistance to the herbicide appeared, it has been possible to control all weeds in a field without harming the crop.

agriculture, the environment, or for our collective future? These are not easy questions to answer. In many cases the first question is answered with a clear yes. We can do it, at least in the short run, in the present crop-pest environment. The second question often has a clear answer, which is nearly always dependent on the bias of the respondent. The answer does not result from a dispassionate analysis of what we ought to do. It is a passionate opinion.

History and the Future

It is my view that weed scientists would benefit from understanding their past, examining their assumptions in light of their history, and then reflecting on the influence of their perception of the desirability of intensive environmental control on the way agriculture is practiced.

When weed scientists know their history and how it influences their choice of weed control and management techniques they will continue the shift that is presently evident away from the dominant chemical control orientation to weed management programs that include but are not dominated by herbicides. Many will greet a request for historical understanding as a nostalgic plea for the past which rejects and diminishes the achievement of successful weed management programs in most crops. Science-based research has developed advanced herbicide technology and improved traditional control techniques that contribute to increased production and profit for farmers while simultaneously diminishing potential harm to the environment and people. Many will also reject the hypothesis that weed science is dominated by herbicides.

I suggest weed scientists should deal with the paradox that while advancing technologies may make present techniques obsolete, they also may revitalize older techniques. For example, computers may (it is far from certain) allow people to regain a measure of individual freedom and abandon assembly line regimentation. They may permit decentralization and dispersal of work places, rather than concentrating them (Friedrich 1984). I do not recommend re-inventing the agricultural wheel. I recommend dispassionate exploration of where we are in weed science and how we got there. I want to question the continued dominance of the technology that permitted the level of weed control that can now be obtained easily and rediscover the obvious that has been abandoned as technology has roared on. Forgotten obvious agricultural things, such as crop rotation, may well be worth rediscovering.

> Yes, what you are saying is all very well in its way. No doubt it would be noble to harden ourselves and do without aspirins and central heating and so forth. But the point is, you see, that nobody seriously wants it. It would mean going back to an agricultural way of life, which means beastly hard work and isn't at all the same thing as playing at gardening. I don't want hard work, you don't want hard work—nobody wants it who knows what it means. You only talk as you do because you've never done a day's work in your life, etc., etc.
>
> Orwell 1937

Orwell, a socialist, was commenting on the socialist critique of progress. He was criticizing what he thought was bad about socialism. His comment was directed toward those who were hostile to a machine-dominated civilization. His words sound like what those who espouse the conventional wisdom in the agricultural community may say to one who advocates rediscovering and exploiting the value of techniques advanced technology has diminished: crop rotation, cultivation, careful land preparation, and multi-cultural crop environments. These techniques should be examined while simultaneously advancing current, excellent work on chemical control, the role of genetic modification, biological weed control, and integration of methods. The conventional wisdom holds that old techniques have not been abandoned, only added to. This must be questioned.

No one wants to return to early twentieth century agriculture. But we should not abandon what we once knew, just because it is old. New weed management methods may offer tremendous advantages and demand solutions to the serious problems that inevitably occur. It has been an advantage for users and developers of agricultural technology that the cost of many problems have been externalized. The best, new or old agricultural technology should solve problems and contribute to solutions of other problems. Undesirable technology solves problems and creates a series of new problems (Berry 1981). For example, the use of some herbicides provides good weed control at reasonable cost. The same herbicides can pollute water and harm people or other species—costs that are externalized. There is no scientific or social requirement that every technological innovation, every new herbicide, and every dream of technological achievement must be indulged. Some technological achievements are becoming proscribed because of their real disadvantages; and others because of their perceived disadvantages. Pesticides have faced serious regulatory challenges and the public is, at best, uncertain about them. Therefore, their continued development may be further proscribed and unquestioning allegiance and advocacy will deter continued agricultural development and inhibit weed science's continued progress.

My opinion is that many weed scientists do not appreciate the extent of their collective commitment to herbicides as the best solution to weed problems. For example, eight weed science textbooks published in the 1960s, 1970s and 1980s devoted an average of nearly 50% (ranged 31–60) to use of herbicides for weed control, herbicide mode-of-action, and herbicide-environment interactions. The emphasis of six textbooks published in the 1990s was nearly identical (range 34–71). Weed science journals show a similar trend. In Weed Science from 2005 to 2010, 34% of the papers dealt with some aspect of herbicides, and it is significant and encouraging that 38% were devoted to weed biology and ecology. For the same years, 71% of the papers in Weed Technology were devoted to some aspect of herbicides, only 12% dealt with weed biology/ecology. The journal's defined purposes account for the difference.

Papers presented at annual meetings of the Weed Science Society of America and the Western Society of Weed Science (Table 1.2) affirm the trend. Both societies might be justifiably accused of being herbicide rather than the weed science societies. Discussions and thoughts among weed scientists are dominated by herbicides not by weeds. It is more accurate to say they are dominated by weed control and one major technology to achieve it, although the rising emphasis on weed biology and ecology indicates change.

> Despite much propaganda to the contrary. there can never be any serious question that pest control chemicals and food-additive chemicals are essential to adequate food production, manufacture, marketing and storage, yet without continuing surveillance and intelligent control some of those that persist in our food stuffs could at times conceivably endanger the public health.
>
> Anonymous 1983

Weed scientists and the great majority of pesticide scientists would agree that there can never be any serious question that pesticide use is essential to adequate food production. Those who readily agree with the claim of essentiality in the quote above may not notice the paradox of combination with danger. There is much propaganda (a euphemism for misinformation or false, non-scientific, emotional critiques) against pesticides and their use. Advocates assume continued use is essential as the quote maintains. Advocates of essentiality believe questions will be answered by continued application of the scientific method by competent scientists. It is accepted, by all parties, that without continuing surveillance and intelligent control, public health could be endangered. Advocates assume that scientific research will answer the serious questions that arise. There is trust that our extensive (some advocates say oppressive) government regulatory scheme will be intelligent and do the necessary surveillance to avoid problems that could, but it is hoped, won't occur. Advocates thereby make the error Berry (1977) accuses them of: "Agriculture experts and agribusinessmen are free to believe that their system works because they have accepted a convention which makes external and therefore, irrelevant, all evidence that it does not work. External questions are not asked or not heard, much less answered."

When there is a finite possibility that some pesticides could persist in food and "could at times conceivably endanger public health" (Anonymous 1983), that possibility raises a serious question, the validity of which cannot be denied by invoking the essentiality of herbicides or any other pesticides. A preponderance of the public has a poor or no understanding of the agricultural view that pesticides are essential to maintenance of food production. To suggest that this justifies concluding that the public cannot and should not seek answers to questions about their essentiality is irresponsible. The public's quest need not result in acceptance of their doubt. Discussion, not dismissal by either side, must follow. Suggesting that the science-based development process and intelligent users are sufficient to prevent all potential problems flies in the face of common sense. The public is regularly notified of fallible technology and weed science has not escaped public scrutiny. To wit:

- The mid-1960s controversy over the real and suspected hazards of 2,4,5-T a component of Agent Orange used in Operation Ranch Hand, a vegetation control program during the Vietnam war. It was the first major public debate that challenged the intellectual foundation of weed science and its dependence on herbicides. (See Chap. 2)

- On December 3, 1984 a poisonous cloud of methyl isocyanate, used in the manufacture of pesticides, escaped from Union Carbide's plant in Bhopal, India killing 14,000 and permanently injuring 30,000 people.
- Pesticide poisoning–no one knows for sure, but it is estimated that between one and 5 million cases of pesticide poisoning occur every year in the world, resulting in 20,000 deaths. Developing countries use 25% of pesticides, but experience 99% of the deaths. (Goldman 2004)
- On April 26, 1986 unit four at the nuclear power plant in Chernobyl, Ukraine exploded causing approximately 6,000 deaths and injuring more than 30,000 people.
- The less serious US nuclear plant problem in 1979 at Three Mile Island.
- Planes that come apart in the air, don't stay in the air, or collide when in the air. November 5, 2010 the explosion of one engine of an Australian Airlines A-380 forced it to make an unscheduled landing in Singapore.
- January 28, 1986, 73 seconds into its flight the space ship-Challenger-exploded, killing seven astronauts.
- The 1989 spill of 11 million gallons of oil from the Exxon Valdez.
- In 2005 Dell recalled 4.1 million notebook computers when it was discovered that their batteries were a fire risk.
- The failure of over 50 levees in August 2005 in and around New Orleans, LA during Hurricane Katrina.
- The collapse of the eight lane, steel-truss bridge on Interstate 35 over the Mississippi River on the evening of August 1, 2007.
- On August 10, 2010 the Deepwater Horizon drilling platform caught fire in the Gulf of Mexico. There have been several other oil spills.
- The huge car recall by Toyota in 2010.
- There were few immediate, reasonable solutions when weed resistance developed in glyphosate-resistant cotton and other crops used by many US farmers.
- The debate within and outside the agricultural community over the risks and ultimate beneficiaries of genetic modification of crops has raised legitimate economic, social and biological concern. The concern for weed science is the widespread adoption of herbicide resistant technology.

The evident fallibility of some technologies and the public debate they engender have a common theme. In each case, highly skilled professionals ignored conspicuous warnings of potential danger and pushed the technology beyond its limits (Evan and Manion 2002, p. 4). The problem lies in the union of the technology with human fallibility. Continued profession of faith in technology and statements that more will cure present problems are not accepted by an increasingly dubious public. If we don't learn from the past, it will be another case of exhausting the usefulness of new technology and waiting for an even-newer one to rescue us. Weed scientists profess appropriate faith, justified by experience and continued success, in their technology, but fail to recognize that the general public sees all technology as related. It is difficult for scientists to understand why the public is so skeptical or to appreciate the public's perception (perhaps the reality) of the awkwardness and inefficiency of institutions designed to regulate technology. Perhaps all technologists can be accused as Evan and Manion claim of ignoring conspicuous warnings of potential danger and pushing their technology beyond its limits.

Weed scientists should not dismiss critics with the epithet—emotional propagandists, environmental wackos, or simply ignorant people. Serious, intelligent questions should be listened to, their origin understood, and then addressed completely and responsibly. We must persevere when correct, yield when wrong, and know how to choose which course to follow. We must also recognize that some questions cannot be answered once and for all. Questions about environmental and human safety will persist because they are inherent in technology. They must be addressed frequently. The best answers may be from those that recognize the nature of the question and its origin in real human concerns that people share. We should not reject the question and thereby patronize the questioner.

The public regards science as a source of facts rather than a process. They do not believe the claim that only the initiated can really understand all the facts required to make good, wise decisions. The public is dismayed when scientists who, it is assumed, should know cannot agree on what the facts are, what they mean, or what ought to be done.[4] How can scientists and technologists expect us to believe their ways will lead us out of the wilderness to salvation when after several decades of spectacular achievements the American public is more anxious, feels less safe, lives in a more polluted environment and a world that is not at peace? We live in an age of anxiety and although we enjoy the marvelous benefits of technological achievements, not the least of which is abundant, safe food, the achievements often make us more, not less, anxious about the world science has created. Science and technology are essential to future planning and key to continued development and progress. However, while they are necessary they alone are not sufficient to solve the complex problems our society faces.

The pharmaceutical industry has been accused of conditioning doctors and patients to believe that the human body needs continued medical supervision and drug treatments to stay healthy (Capra 1982). Many believe the agri-chemical industry is similarly culpable of conditioning farmers to believe that soils and crops need pesticides, whose application should be supervised by properly trained agricultural technicians. The claim is that without proper pest management no crop can reach maximum productivity. In both cases, these practices have (not always) seriously disrupted the natural balance of living systems and thus generate unease

---

[4] When Senator Edmund Muskie of Maine was working to develop the 1970 US Clean Air Act he pleaded that the scientific information gathered in support of the act be provided by scientists with only one hand. Then he would no longer hear scientists say "on the other hand." Senator Muskie wanted what many want—indisputable scientific facts in support of the legislation. Unfortunately, that is not how science proceeds.

and numerous diseases (Capra). Each year, around 5 billion pounds of pesticides[5] with a market value of $32 billion, are used on the planet's crops. In 2007, the US pesticide market was valued at $12.5 billion. The public learns, usually from the media, that in the US, nearly one in ten pesticides may be capable of causing cancer. It, as mentioned above, is a risk that scares people, but it may not be true.

Pesticide use has expanded dramatically since the discovery of DDT in 1939. In February 2001, 1,200 active ingredients were used in 17,270 pesticides registered (approved for use) by the US/EPA. An additional 2,530 products were registered for special needs.[6] Donaldson et al. (2002) reported that approximately 16,000 pesticides based on approximately 600 active ingredients were registered with the EPA in 2002. About 80% was used in agriculture (Donaldson et al.).

In 1984, Pearson estimated 1.5–2 million people, in the world's developing countries, suffered acute pesticide poisoning every year. Twenty years later, Goldman (2004) estimated 1–5 million poisonings and a few thousand fatalities each year. Jeyaratnam (1990) reported that there may be 1–3 million cases of pesticide poisoning in the world each year, and based on data from the World Health Organization (WHO), there may be 220,000 deaths, nearly all in developing countries. Bronstein et al. (2009)estimated that 10,000–20,000 deaths occur annually, primarily in developing countries. Nearly all poisonings occur where people don't know how to use pesticides properly, do not have access to protective clothing or washing facilities and there is insufficient enforcement of safety standards that may not exist and, if they do, may be of no help to workers who can't read. An example, relevant to this essay, is paraquat poisonings in Malaysia. Seventy three percent were due to suicide. Fifteen percent to accidents and slightly less than 1% to occupational accidents. The best summary of the data on pesticide poisoning is nobody knows for sure; all data are estimates. The benefits of pesticide use are great, but there are large social and environmental costs. Human poisoning is clearly the highest price paid for expanded pesticide use (Mellor and Adams 1984). Pesticide intoxication is a real, but not widely acknowledged public health problem to which the United States contributes by exporting about 1.7 billion pounds of pesticides each year, including legal export of 27 million pounds of products banned for use in the US (Smith, et al. 2008).

A complex mixture of facts, accusations, and well-documented reports of malpractice and environmental damage or human injury serve to harden public attitudes about risk and increase the burden of government regulation, which the pesticide industry blames for hobbling progress of its research based program (Anonymous 1987). The world's major pesticide producers spent about $410 million on research and development in 1985 which was about 8.9% of farm sales ($4.6 billion) and 6.2% of total sales ($6.6 billion). The 1960s thalidomide tragedy

---

[5] Throughout this essay, the generic category pesticide is used. To the best of my knowledge, data on herbicides alone are not available.

[6] Data provided by Claire Gesalman, Chief, Communication Services Branch Office of Pesticide Programs, EPA. February 2011.

and other problems have reinforced the public belief that drug companies are not staffed exclusively by saintly scientists and this increased demands for legislation to prevent future tragedies (Anonymous 1987). As scientifically inappropriate as it may be, that tragedy reflected negatively on all chemical manufacturers. Public awareness of pesticide misuse and subsequent problems, even if wrong, inform public opinion. The public assumes other tragedies will follow.

Perception of Risk

People are willing to accept higher risks from activities perceived to be highly beneficial and where one feels in control. We drive cars willingly although the risk of an accident is presumed to be high. Annually 40,000 people die in car accidents in the US Levitt and Dubner (2009 p. 151) acknowledge that many more people die in car compared to airplane accidents each year. A primary reason is that people spend a lot more time driving cars than in airplanes. Levitt and Dubner advert that the per-hour death rate for driving and flying is about equal. We accept the higher risk of driving because we feel in control, whereas we do not when flying.

The possible hazard of pesticides in the diet of the average United States citizen is quite low. It Is roughly equivalent to that of the chloroform (a known carcinogen) actually present (about 83 ppb) in one glass of average US tap water. It is insignificant compared to natural carcinogens in our diet (Ames 1985). We know that coffee contains the known natural carcinogens, hydrogen peroxide and methylglyoxal at about 4000 ppb while cola drinks contain the known carcinogen formaldehyde at 7,900 ppb. Beer contains known carcinogens and alcohol consumption can cause human cancer. Perhaps most shocking, milk, with its high percentage of fat, may be implicated in human breast and colon cancer both associated with high fat consumption. These facts are used by pesticide advocates to support pesticide use which, they maintain, is less hazardous than many common routinely accepted risky human activities (e.g., skiing, mountain climbing, and sky diving).

Research by Slovic (1987) suggests that there is wisdom and error in public attitudes and perceptions. The "basic conceptualization of risk is much richer than that of experts and reflects legitimate concerns that are typically omitted from expert risk assessments." Slovic's data show that pesticides are unknown hazards, by which he means they are not observable to those who may be exposed, their effects may be delayed, and the totality of risk may be unknown. He also categorizes them as dreaded, often fatal, not equitably distributed, having a high probability of risk to future generations, and possible effects are not easily mitigated once released in the environment. A dread risk may be one which is increasing with time, is involuntary, and regarded as outrageous. Hazards judged to be voluntary (driving a car, riding a bicycle, smoking, using a home lawnmower) tend to be judged as personally controllable. Pesticides in Slovic's opinion are involuntary risks which are unobservable, regarded as definite hazards, perceived to have delayed harmful manifestations and are, therefore, regarded as uncontrollable. Pesticide advocates often respond to this kind of risk analysis by requesting a rational look at all the data and a complete examination of risk/benefit information. Advocates find it difficult to accept that the most reasonable assumption about people's behavior is that there is some method in any apparent madness (Fischoff et al. 1981). People may be wrong but they are not stupid. Weed scientists rely heavily on logical positivism and their intuition which they are sure is based on objective evaluation of all important data.

Pesticides also suffer from the negative image shared by large companies whose activities are suspect perhaps only because they are large, not personally controllable, and remote from common experience. Pesticide companies are accused of selling dangerous, unnecessary products (e.g., export of banned pesticides) in developing countries that do not or cannot regulate markets (Bull 1982; Weir and Schapiro 1981). Companies plead not guilty because they are providing products developing countries need to develop. They argue it would be morally unacceptable to deny access to potentially beneficial technology and to presume they can or should do what countries ought to do—govern local use with adequate regulation and intelligent supervision. The industry is also accused of over charging for products and using people in developing countries as guinea pigs to test products that cannot yet be sold, may never be sold, or can no longer be sold in developed countries. However, in terms of the industry's image, it matters little whether these accusations are true. They are believed (Anonymous 1987). There are many parallels between the pesticide dilemma and the nuclear power industry's dilemma. Both are widely feared, both suffer from a nearly universal misconception about what they are and can do. Both have generally acknowledged dangers, and advocates of both say that unnecessary and increasing governmental regulation is hampering continued development. They also share wide spread public disagreement over present policy and interpretation of danger.

What should be done when experts disagree even about problem definition? In view of the fact that the public believes the pesticide chemical industry is guilty as charged, Prior (1962) questions whether today's scientists can maintain allegiance to the ethical imperative of searching out the truth. He suggests that scientists should seek truth but have a unique moral obligation to reflect on implications of their discoveries, to inform others, and to search for answers to new questions created by science's undeniable effects on public issues. Scientists express their values when they choose what problems to study and the methods used to study them. Many agricultural scientists resist this challenge and do not speak out in the public arena about their work and its meaning because they are reluctant to be perceived as reaching beyond their area of technical expertise—their island. They see public debate as inevitably political and as an unknown, irresponsible arena. However, policy makers will make decisions based on poor science or with a poor understanding of science unless scientists are willing to speak individually and collectively. Thus, agricultural scientists, including weed scientists, are not value free and certainly are not morally superior. We succumb to the vices and possess the virtues of ordinary people in about the same measure and the public knows it. We are viewed as performing a function that can (and should) be judged in its social context and evaluated according to the means used and the ends served.

Scientists have different opinions about scientific matters. Indeed it is the essence of science that it be a falsifiable activity and debate within the scientific community is essential to verification. The scientist's judgment is influenced by political and other biases. For example, the result of much science is not publicly known because it is conducted in isolation from public scrutiny. Who reads scientific journals? The source of funding, proprietary nature, and potential patentability of research are not revealed even when results are published. The virtues of science and the scientific method are effective only where the matters at issue are purely scientific (Prior). Pesticide and nuclear power issues, inevitably include questions and answers that are not scientific and therefore are, or ought to be, publicly debatable. Determining what issues are purely scientific is not easy. Prior points out the facts about nuclear fallout were known to both sides of the debate and radiation dangers were agreed upon long ago. Differences arose about whether dangers all agreed were real, slight or great, and if inherent risks were presumed to be necessary and slight or unnecessary and great.

The same differences occur in arguments about pesticides. The established facts about their physical and chemical properties, their activity and selectivity, their use and misuse, and their hazards have been reasonably well established for some time. The disagreement is over their essentiality, especially in agriculture, and whether risks attendant on their use are necessary and slight or unnecessary and great. As a consequence, pesticide arguments have shifted to issues that bear only indirectly on scientific answers (Prior). It is argued that continued reliance on pesticides as the front line of defense against pest-caused reductions of crop yield or possible injury to human health is necessary to maintain the quantity and quality of the food supply. Although there are risks, the argument goes, acceptance is common. Possible increases in cancer or environmental pollution from continued or even expanded pesticide use is more than offset by the advantage of increased food production and presumed environmental quality in a world where too many people are not fed adequately. Appeals to faith in technological progress and a better future, are offered as scientific arguments. They are not scientific arguments. The decisions they support are not scientific decisions (Prior 1962). The reasoning is by analogy and the argument is a moral one that supports a moral decision: There is hunger in the world, we can help, and ought to do all we can.

Discussions of the necessity of herbicides and other pesticides for continued agricultural progress are not debates about scientific evidence. They are value-laden arguments and must be addressed as such. Scientific evidence sometimes is essential, but just as often it may obscure the real issues and make resolution more difficult. The presence of known carcinogens in drinking water, beer, or milk is interesting, but not immediately relevant especially when presented to demean or ridicule those who oppose pesticide use because of their perception of possible health hazards.

Weed scientists like other scientists must have funds to pursue their work. It is reasonable to ask if scientists seek truth or funding. In his 1961 farewell address, President Eisenhower warned us about the military-industrial complex that had become a fulcrum on the American political scene. He also attacked education's role. The university, "the fountainhead of free ideas and scientific discovery, has experienced a revolution in the conduct of research... A governmental contract becomes virtually a substitute for intellectual curiosity." If that is the public's perception of the academic community is it any wonder that our arguments in favor of, and our demonstrated allegiance to, a particular production technology and its risks are greeted with suspicion about our true motives?

The Problem

Weed scientists may be akin to the Sorcerer's apprentice, unable to control the forces unleashed. Perhaps we are modern Captain Ahabs, all our methods are sane but our goal is mad. How much control can we logically achieve? Can weed scientists achieve almost any degree of control they want and what should they want? Do weed scientists suffer from the same excessive faith in their expertise (Berry 1977; Laski 1930) that all experts may have? Do they neglect all evidence that does not come from within the ranks? Do they fail to see the values inherent in their decisions and the actions taken based on those decisions? Do they confuse the importance of facts about herbicides and weed control with the importance of what they propose to do with facts they choose to accept? If one or more of these things is true, what should be done by whom? Some may argue that these attitudes and the results do not define a problem. I disagree. The best long-term solution is to change attitudes and values. Environmental education which justifies rights for rocks (Nash 1977) animals (Singer 2002; Rollin 1995) and trees (Stone 1974) tends to discredit the confidence that a technological fix exists for every problem.

The quest for environmental quality and long-term stability has run counter to the growth oriented American ethos. The US has been growth oriented since its founding. It is appropriate to explore our dedication to economic growth and the world-wide imperialism it generated and continues to foster. Is it sustainable? Daly (1996) p. 7) argues that sustainable growth is a dangerous oxymoron. Sustainable growth has been used as a synonym for sustainable development. It is impossible. Growth, a quantitative increase, is impossible to sustain. Children grow and develop simultaneously, cease growing, and continue to develop. The earth develops but cannot grow (Daly p. 167). Development is qualitative change (Daly p. 167). Quantitative change has been regarded as the only solution to poverty, something we ought to do. Qualitative change, sustainable development, does not dismiss the benefits of growth but it must occur within the limits of the earth's carrying capacity. Development considers environmental and ecological limits. The dominant growth ethic has demanded environmental rapaciousness and wholesale neglect of the environmental consequences of economic development (growth). Collective world progress, and perhaps real world peace, await substitution of ethical for economic criteria in the calculation of human and technological effects on the world and on others (Nash 1977; Singer 2002). In our quest for continued growth, we abuse land and, therefore, ourselves because we regard land as a commodity which belongs to us rather than as something to which we belong and care for temporarily. Leopold (1966) told us that "a thing is right when it tends to preserve the integrity, stability, and beauty of the biotic community. It is wrong when it tends otherwise."

Agricultural scientists may know Leopold's words but have not learned his lesson well and do not use his ethical standard when economically based control technology is developed. Values of the right to existence of the unknown or inanimate are among the latest moral acquisitions and they are neither commonly discussed in weed science nor are they pervasive in agriculture. Agriculturalists are secure in the correctness of their views because they constantly share them with each other. When they speak outside familiar circles they often become frustrated. It is easy to blame the recipient of a message for an inability to understand (Hoffmann 1987). Perhaps recipients are scientifically illiterate or so committed to their own narrow view that they either cannot hear or choose to ignore the clear logic of the agricultural view. It is much more comfortable to speak to those who understand our language and know the wisdom of our words.

Issues of Fact and Value

Separating issues of fact from issues of value is fundamental to intellectual hygiene (Fischoff et al. 1981). The pesticide debate includes opponents willing, perhaps eager, to give quick, easy definitive answers to complicated questions. These answers often go beyond available data and pesticide advocates quickly label them as emotional claims. Often they are. But, emotion does not always symbolize error and often carries the debate in spite of an evenhanded scientific approach. Commonly, debaters do not really speak to each other because issues of fact are not distinguished from issues of value (Fischoff et al.). Separation is not always possible because there is no clear distinction. There may be values within facts and vice versa. Beliefs about facts shape values and those values, in turn, shape facts we search for and how we interpret them when we find them. Knowing this abstractly still means all in the debate are bound by their education and traditional ways of defining and studying problems (Fischoff et al.).

Scientists pride themselves on objectivity but it frequently has a subjective dimension that assumes they are addressing the right problem, correctly. They assume issues of fact and value are correctly sorted. Acquisition of objectivity is further complicated by definition of the right problem and acceptable solutions. One hopes the best solutions will be defined as those in society's best interests, not by what is in the best or exclusive interest of one party to the debate (Fischoff et al.). Debates about pesticides will continue. It is my hope that weed scientists will see that it should not be their purpose to win. Attempts to win are doomed to fail. Winning means someone else loses and in value issues proving someone else wrong does not prove your side right. It is our burden to define risks and benefits and then discuss acceptability of risk in light of benefits. This is not an easy task and one many scientists are ill equipped for. There are no universally acceptable risks and the lowest risk option may not be best for all (Fischoff et al.).

Issue Resolution

Resolution of pesticide issues is complicated by at least three major dilemmas (Farrell 1983). The *first* is a lack of complete scientific evidence concerning basic environmental and human health implications of pesticide use. We know a great deal about how pesticides act and how they can be used to maximize their activity and selectivity. With few exceptions, too little is known about what many consider to be more important questions. The complete fate of a pesticide after it leaves its point of application cannot be described, except in general terms. It is agreed that transport occurs, but we don't know how far, in what form or what the complete path of degradation/metabolism is. Although a great deal is known, we do not know the complete fate of an applied molecule and its metabolites. Groundwater contamination by pesticides is a major public and scientific issue and is being addressed competently in many research laboratories. It has always been of concern to pesticide developers, but those involved in pesticide development (not just the original manufacturer) did not act swiftly to prevent problems although they now are working vigorously to solve identified problems. The *second* dilemma is the difficulty, and perhaps the inability, to evaluate fully the social and environmental costs of pesticides on human health, non-target organisms, and on water, soil, and air. The dilemma's questions have been neglected in the quest for increased sales and efficient pest control. A *third* dilemma, complicating final resolution of these issues, is the inadequacy of institutions designed to allocate costs and resolve issues even after adequate problem identification has been achieved.

The public has come to recognize that each advice giver has particular strengths and weaknesses and should not be allowed, independent of other advice givers, to make societal decisions (Fischoff et al. 1981). No one view can fully present or assess the effect of pesticides at their point of use and predict their future effects. Risk management is too complex and too risky to leave to experts. Experts are necessary because of their expertise, but it is not sufficient to allow them to make all final decisions and assume they will make the best decisions for all. Pesticide use is coming to be recognized as more than just a crop production decision that a farmer and advisor can make. There are too many ramifications that go beyond a site and intended use. Because of their expertise, weed scientists and others in agriculture must participate in use decisions but should not presume they should be made solely within the context of crop production. Weed scientists and agricultural scientists have essential information and insights to offer but they may not have the correct answer to all questions that lie outside their expertise. Admitting relative ignorance may make science and scientists more credible and allow them to continue to make outstanding contributions through enhancing food production and environmental quality. One way to do this is to continue development of more systematic approaches to weed management rather than relying on weed control to solve all problems. We would not tolerate for long a medical or dental practitioner who cured a disease once diagnosed, but failed to emphasize disease prevention. Most people intuitively understand that it frequently is easier, healthier, less expensive and more pleasant to prevent a disease than to cure it once it has attacked. Children in the developed world routinely are immunized to prevent contraction and spread of contagious diseases because most parents recognize the health consequences of serious diseases of childhood many of which are nearly eradicated from the developed world due to development and use of vaccination. We do not know how to vaccinate an agricultural field to prevent insect, disease, or weed infestation. However, one thing all pest control specialists can be justifiably criticized for is stressing control technology while giving only minor attention to preventive technology. Fryer (1985), former Director of the Weed Research Organization, Oxford, UK, agreed when he pointed out that herbicides have provided a powerful tool for weed control, and their great efficiency has enabled intensive agriculture to develop. It is not appreciated that emphasis on chemical control technology has generated new weed problems which seem to require more complex and intensive herbicide treatments. In this system, which dominates developed country agriculture, reliance on herbicides has become complete. Fryer asks the following questions which concluded the 1991 version of this essay.

(1) Has reliance on empirical use of herbicides as a substitute for the mechanical and cultural methods of weed control formerly practiced gone too far?
(2) Might not greater efficiency and cost effectiveness in crop production be obtained if cultural and chemical methods of weed control were to be harnessed more positively to work together?
(3) By a better understanding, through research, of the weeds themselves, of their ecology, of the constraints they impose on crop production, and of the factors influencing population trends in particular cropping regimes, could not a more rational approach to weed control result in greater economy in the use of chemicals and improved environmental quality or, where herbicides cannot be used, in more efficient weed control, higher yields and better use of humans and other resources?

He proposed the answer to each question is yes and therefore greater emphasis should be given to weed management as opposed to control. Management will force practitioners to redefine their practice. Words and the definitions we create from them are powerful determinants of how we act and what we propose to do. Weed management, Fryer suggests, will help us move away from the "empirical and often uninformed application of physical or chemical techniques to control weeds toward the rational deployment" of all available technology to provide systematic management of weeds in all situations.

Among several excellent examples of integration of techniques, Fryer cites work by Cussans (1976) who calculated that if annual herbicide use in continuous spring barley gave an 80% reduction of viable wild oats (*Avena*) seed, the soil population would continue to rise if farmers tine-cultivated soon after barley harvest. If tine-cultivation was omitted and plowing delayed until December, a slow decline in the soil seed bank occurred and the problem diminished.

References

Ames BN (1985) Testimony to California senate committee on toxics and public safety management. Sacramento, CA, November 11

Anonymous (1983) Preface. Residue Rev 85:VIII

Anonymous (1987) Pharmaceuticals: Harder going. The Economist Feb. 7:3–6, 9–20, 13–18

Bailey R (Ed) (1995) The true state of the planet. The Free Press, New York, p 472

Berry W (1977) The unsettling of America–culture and agriculture. Avon Books, New York, p 228

Berry W (1981) The gift of good land: further essays cultural and agricultural. North Point Press, San Francisco 280 pp

Black JN (1970) The dominion of man: the search for ecological responsibility. Edinburgh University Press, Edinburgh, Scotland, p 169

Bronstein AC, Spyker DA, Cantilon LR, Green JL, Rumack BH, Giffen SL (2009) 2008 annual report of the American association of poison control centers. National poison data system (NPSS), 26th annual report. Clin Toxicol 47:911–1084

Brown LR (2004) Outgrowing the earth: the food security challenge in the age of falling water tables and rising temperatures. W.W Norton & Company, New York, p 239

Bull D (1982) A growing problem–pesticides and the third world poor. Oxfam, Oxford, p 192

Capra F (1982) The turning point. Bantam Books, New York, pp 87, 252–253.

Cussans GW (1976) Population dynamics of wild oats in relation to systematic control. Report, ARC Weed Res. Org., Oxford, UK, 1974–1975, pp 47–56.

Daly HE (1996) Beyond growth: the economics of sustainable development. Beacon Press, Boston, p 253

Donaldson D, T Kiely, A Grube (2002) Pesticides industry sales and usage. 1998 and 1999 market estimates. Washington, DC, US environmental protection agency, Report no. EPA-733-R-02-001

Durning AT (1996) This place on earth: home and the practice of permanence. Sasquatch Books, Seattle, p 326

Ehrlich PR, Ehrlich AH (1996) Betrayal of science and reason: how anti-environmental rhetoric threatens our future. Island Press, Washington, p 335

Evan WM, Manion M (2002) Minding the machines: preventing technological disasters. Prentice Hall, Upper Saddle River, NJ, p 485

Farrell KR (1983) Critical choices for natural resources. Chap. 1. In: Rosenblum JW (ed) Agriculture in the twenty first century. Wiley Interscience, New York

Fischoff B, Lichtenstein S, Slovic P, Derby SL, Keeney RL (1981) Acceptable risk. Cambridge University Press, Cambridge, 185 pp

Friedrich O (1984) Adieu to the Pneu. TIME essay, April 30, p 82

Fryer JD (1985) Recent research on weed management: new light on an old practice. Chap. 9. In: WW Fletcher (ed) Recent advances in weed research. The Gresham Press, Old Working, Surrey

Galbraith JK (1958) The affluent society. Houghton Mifflin, Boston, 368 pp

Goldman L (2004) Childhood pesticide poisoning: information for advocacy and action, United Nations environment programme. Châtelaine, Switzerland, p 37

Gore A (2006) An inconvenient truth. Rodale, Emmaus, PA, p 325

Hoffmann R (1987) Plainly speaking. Am Sci 75:418–420

Jackson W (1980) New roots for agriculture. Friends of the Earth, San Francisco and The Land Institute, Salina, KS, 155 pp

Jeyaratnam J (1990) Acute pesticide poisoning: a major global health problem. World Health Stat Q 43(3):139–144

Kirschenmann F (2010) Imagining resilience. Leopold Lett 22(3):5

Laski H (1930) Diderot: homage to a genius. Harper's Mag 162:597–606

Leopold A (1966) A sand county Almanac. Ballantine Books, New York, p 295

Levitt SD, Dubner SJ (2009) Freakonomics: a rouge economist explores the hidden side of everything. Harper -Perennial, New York, p 315

Lomborg B (1998) The skeptical optimist: measuring the real state of the world. Cambridge University Press, Cambridge, p 515

Mayer A, Mayer J (1974) Agriculture, the Island empire. Daedalus 103:83–95

McKibben B (2003) Enough: staying human in an engineered age. Henry Holt and Company, New York, p 271

Mellor JW, Adams RH Jr (1984) Feeding the underdeveloped world. Chem Eng News 23:32–39

Nash R (1977) Do rocks have rights? The Center Magazine, Nov. Dec., pp 2–12

Orwell G (1937) The road to Wigan Pier. The Left Book Club, UK, pp 209–210

Prior EM (1962) Science and the humanities. Northwestern University Press, Evanston, p 124

Ridley M (2010) The rational optimist: how prosperity evolves. Harper Collins Publishers, New York, p 438

Rollin BE (1995) Farm animal welfare: social, bioethical, and research issues. Iowa State University press, Ames, p 168

Simon J (1981) The ultimate resource. Princeton University Press, Princeton, p 415

Simon J (1995) The state of humanity. Blackwell, Cambridge, p 694

Simon J, Khan H (1984) The resourceful earth-a response to global 2000. Basil Blackwell, New York, p 585

Singer P (2002) Animal liberation. Harper Collins publishers, New York, p 324

Slovic P (1987) Perception of risk. Science 236:280–285

Smil V (2000) Feeding the world: a challenge for the twenty-first century. The MIT press, Cambridge, p 360

Smith C, Kerr K, Sadripour A (2008) Pesticide exports from US ports, 2001–2003. Int J Occup Environ Health 14(3):176–186

Stone CD (1974) Should trees have standing? Toward legal rights for natural objects. William Kaufmann, Inc, Los Altos, p 102

Weir D, Schapiro M (1981) Circle of poison-pesticides and people in a hungry world. Inst Food Dev Policy, San Francisco 96

White L Jr (1968) The dynamo and the virgin reconsidered: essays in the dynamism of western culture. The MIT Press, Cambridge 186

Pearson CS (1984) What has to be done to prevent more Bhopals? The Washington Post, Dec. 9, 1984

Zimdahl RL (1978) The pesticide paradigm. Bull Entomol Soc Am 24:357–360

Chapter 5
The Future

But you who seek to give and merit Fame, And justly bear a Critick's noble Name, Be sure your self and your own Reach to know. How far your Genius, Taste, and Learning go; Launch not beyond your Depth, but be discreet, And mark that Point where Sense and Dulness meet.

An Essay on Criticism, Part 1
Alexander Pope, 1709

One of the papers published in Weed Technology as a follow-up to the 1993 symposium (Zimdahl 1994) suggested that weed scientists should come to "a reasonable accommodation between a productive and protected environment." Five ideas about weed science were briefly explored.

- In its early days weed science might have been more properly called herbicide science (Thill et al. 1991; Merrigan 1993).[1]
- Reliance on the use of herbicides to control weeds has become too great.
- Weed science will benefit as integrated, sustainable weed management systems are developed.
- Weed science is an economically dominated, technical specialty divorced from its basic science—botany.
- Weed science is moving toward domination by population biology and ecology.

The paper went on to claim that "in earlier times, things were dear and people were cheap." Agriculture included many destructive practices e.g., plowing and the dust bowl (Montgomery 2007; Egan 2006). Sound ecologically-based practices including crop rotation, long-term crops—pasture, limited use of external inputs, etc. were often the way things were done, probably because they were the only way to do things. As agricultural technology, (including herbicides) developed, things became cheap and people dear and "ecology, evolution, carrying capacity, limiting resources, and limits to growth could be and were ignored."

Bridges (1994) discussed the economic, aesthetic, and health effects of weeds and presented data on the economic effects of weeds in the US economy. His purpose was to demonstrate present effects and costs of weeds on human endeavors. In his view, weeds were important and their management would continue to be important as long as humans altered and created habitats for the practice of agriculture.

---

[1] Dr. Merrigan is Deputy Secretary of the US Department of Agriculture.

R. L. Zimdahl, *Weed Science: A Plea for Thought*—Revisited, SpringerBriefs in Agriculture, DOI: 10.1007/978-94-007-2088-6_5, © Robert L. Zimdahl 2012

Coble, Hess, Holt, and Wyse presented complementary, slightly different versions of the future of weed science. Holt (1994) suggested that "with a foundation of knowledge in basic weed biology, alternatives to herbicides can be made available in the future." Her suggestion was influenced by the fact that the, largely urban, US public was (79% in 2010) concerned about the effects of agricultural practices on food safety, environmental quality, sustained productivity, and perhaps maintenance of farm communities. She thought that weed science was in transition from a largely technological to a broadly fundamental discipline. In 1994, few alternatives to herbicides and tillage were available.

Coble (1994), Hess (1994), and Wyse (1994) offered specific recommendations for appropriate changes in weed science research. Coble thought development of long-term sustainable weed management depended on the ability to predict future consequences. In his view, there was insufficient knowledge of most aspects of weed biology (e.g., weed seed population dynamics, species shifts over time, and adaptation to selection pressures). He strongly recommended development of economic thresholds. His view is consistent with the view that the justification for application of weed management techniques should not be— "I have always done it this way," or "just in case." Weed management should be done only when positive economic beneficial results can be predicted with a high degree of confidence.

Hess suggested the need for research in seven areas that focused on understanding the effects of weeds in several habitats and efficient management of weeds with minimal environmental effect. Four of his suggestions emphasize research on herbicides to:

- Reduce herbicide contamination of surface and groundwater,
- Modify [herbicide] application technology to increase weed management efficiency,
- Improve understanding and management of weed resistance to herbicides,
- Improve techniques for detecting herbicides in the environment.

His other recommendations:

- Develop a systems approach to understanding the effect of weed management,
- Increase understanding of weed biology,
- Biological and natural product weed control.

Hess' recommendations differed slightly but were similar to those advocated by Wyse, who described, with a bit more specificity, research that, in his view, would develop ecologically-based methods of weed management as a necessary part of sustainable crop production systems. He proposed weed scientists should address six important research areas.

- Develop new crops or adapt present crops that create varying patterns of resource competition, allopathic interference, and soil disturbance that prevent weed proliferation.

- Explore the development of crop rotations that reduce energy, tillage, fertilizer, and pesticide use; but maintain yield and profitability.
- Develop cover/smother plants that suppress weeds without reducing crop yield.
- Develop crop varieties with enhanced weed interference potential.
- Develop new biological weed control techniques.
- Develop simulation models that include biological, economic, and environmental effects of weed management decisions.

The last point is consistent with Coble's recommendation for development of economic thresholds. These distinguished weed scientists offered their thoughts about important, if not necessary, future directions for weed science research. They each included development of ecologically-based methods of weed management that would lead to sustainable crop production systems. A reasonable conclusion of their recommendations is that weed scientists need to become better ecologists, better observers of the environment, and managers or primary participants in programs that include collaboration with other agricultural disciplines. These more inclusive programs should be focused on creation of "sustainable crop production systems that improve or, at a minimum, do not harm the environment, produce healthy and plentiful food, do not harm humans or other species, and support viable rural communities" (Wyse 1994). That is, weed scientists should use their Promethean power to create solutions to weed problems, develop and refine control techniques, apply them when necessary, and be ever watchful for modern plagues similar to those Pandora's curiosity released.

Weed and agricultural scientists have known, in contrast to Eiseley's man on the subway, where they are going. They have been sure of the correctness of their methods and research goals; and, if necessary, justification was easy. The reasons for the journey were known, albeit unexamined. Weed scientists have been sure of the destination, but only rarely have they paused to consider if the journey is taking them where they ought to go. Is the technology leading toward the accepted goals of ecologically-based methods of weed management and sustainable crop production systems advocated by the speakers at the 1993 symposium. Will the current research lead where weed scientists have said they ought to go. It is quite possible that the conclusion after careful thought will be: Yes, weed scientists are doing the right things and will achieve what they said they wanted to achieve. In this case, the debate, the arguments, the discussion will not have been time wasted. Weed scientists will have carefully articulated where they are going and why. The process will logically include examination of the ethical foundation of activities and their associated values.

A comment by the Russian author Leo Tolstoy[2] about art is relevant and similar to the earlier comment by Laski (1930, Chap. 4). Tolstoy urged questioning and debate about the correctness of assumptions. He acknowledged that it is sometimes difficult and painful to do so.

---

[2] Tolstoy, L. 1904. What is Art? The Christian Teaching. Page 274 in Resurrection Vol. II. Translated and Edited by L. Wiener. Boston, MA. Dana Estes & Co. Pub. I found the quote in Dyson, F. 1984. Weapons and Hope. New York, NY. Harper and Row, Pub. p 213

Tolstoy commented:

> I know that the majority of men who not only are considered to be clever, but who really
> are so, who are capable of comprehending the most difficult scientific, mathematical,
> philosophical discussions, are very rarely able to understand the simplest and most
> obvious truth, if it is such that in consequence of it they will have to admit that the opinion
> which they have formed of a subject, at times with great effort–an opinion of which they
> are proud, which they have taught others, on the basis of which they have arranged their
> whole life–that this opinion may be false.

To preserve what is best about modern weed science and identify the abuses
modern technology has wrought on our land, our people and other creatures, and
begin to correct those abuses will require many lifetimes of work (Berry 1999).
To progress, we must consciously bring all of agriculture's many roles
together—productive, profitable, scientific, environmental, economic, social,
political, and moral. It is no longer sufficient to justify all activities on the basis
of increased productivity and profitability for various agricultural enterprises.
Other criteria, many with clear moral foundations, must be included. We live in
a post-industrial, information age society, but we do not and no one ever will
live in a post-agricultural society. All societies have an agricultural foundation
within their borders or they rely on an agricultural foundation elsewhere. Those
of us in agriculture must strive to assure that the foundations are sound and
secure and will remain so for all of humankind. For weed science, this means
analyzing whether the discipline is sufficiently engaged in the research directions
recommended in 1994 by Bridges, Coble, Hess, Holt, and Wyse. Were the
research ideas proposed in the 1993 symposium good? Have they been pursued
and achieved or is there clear progress toward achievement? If the objectives
have not been pursued, there may be several plausible reasons. To wit:

- They were not a consensus view.
- They were reasonable but not considered realistic or achievable and were thus,
  ignored.
- They were reasonable goals but achieving them proved too difficult.
- There were other easy avenues to research funding and professional success.
- Herbicide technology was developing rapidly, offered short-term success, and
  was, or could be immediately applied by farmers.

Many argue that while the recommended objectives are reasonable, it is implicit
that adopting them will halt or diminish technological progress and food production will decline. Then the burden of those who want to change course will be
to decide who will starve, because that is the inevitable outcome. Those engaged in
agriculture will thereby abandon agriculture's primary and morally good mis-
sion—to produce food.

Some of the important, if not necessary, thought, this essay encourages, concerns whether the opinions of their science that have been formed so carefully are
true or false. After debate and extended discussion, weed scientists may decide and
provide reasons that convince others (scientists, the public) that their opinions of
their science, its goals and its destination are correct. Weed science is headed

where it ought to go. However, if the direction and goals are not adequate to the task and cannot be justified, then debate and discussion must continue.

Agricultural scientists have assumed that as long as their research and the resultant technology increased food production and availability, they and the end users were somehow exempt from negotiating and re-negotiating the moral bargain that is the foundation of the modern democratic state (Thompson 1989). It is a moral good to feed people and agriculture does that. Therefore, its practitioners assume that anyone who questions the morality of agriculture or its technology simply does not understand the importance of what is done. It is assumed that researchers are technically capable and that the good results of technology make them morally correct. My plea, my hope, is for the discussion to begin and continue.

As in so many complex areas of modern life, those who are engaged in and those concerned about food production and the negative effects of agricultural practices (e.g., pesticide pollution, soil erosion, and water depletion) yearn for objective truth that settles issues. Many think they have found it and are willing to describe it for the rest of us. However, it often becomes evident that those who are sure they have objective, publicly verifiable, universal truths probably don't; and, moreover, they may not really understand the question or the reasons people are concerned. This essay is not an attempt to present objective truth, as I hope will immediately be seen. The essay suggests that it is reasonable to ask if the weed science community has "overcome the paralysis of the pesticide paradigm and conceive a weed science research program that addresses both society's perception of safety and the scientific community's perception of risk" (Naegele 1993). It is clear that weed scientists and the discipline's research emphases have changed their focus and goals. The essay's primary plea is to ask weed scientists to think about whether the science's direction and goals are acceptable or need to be modified, and provide reasons modification is or is not necessary. Weed scientists are the immediate audience, but the message is applicable to a wider agricultural audience. My plea is not intended to demean the great accomplishments of weed science. The essay neither concludes with a definitive prescription about where weed science ought to go nor does it judge the path we have or should have followed. It concludes with where it began—with a plea for thoughtful analysis of the path we are on and whether it is taking us where we know we should go. I do not wish to become a tolerated heretic within weed science by appearing to abandon what has been accomplished and those who have accomplished it. Both are to be admired and respected. However, if heresy is my crime and burning at the intellectual stake my punishment, my final plea will be for thought about future directions for weed science, as I go up in smoke.

References

Berry W (1999) In distrust of movements. The Land Report 65 (Fall):3–7. The Land Institute, Salina, KS
Bridges DC (1994) Impact of weeds on human endeavors. Weed Technol 8:392–395
Coble HD (1994) Future directions and needs for weed science research. Weed Technol 8:410–412

Egan T (2006) The worst hard time: the untold story of those who survived the Great American Dust Bowl. Houghton Mifflin Co., New York, 340 pp.

Hess FD (1994) Research needs in weed science. Weed Technol 8:408–409

Holt JS (1994) Impact of weed control on weeds: new problems and research needs. Weed Technol 8:400–402

Laski H (1930) Diderot: Homage to a genius. Harper's Mag 162:597–606

Merrigan KA (1993) There is no such thing as weed science. Proc. Southern Weed Science Society. 43:3–8

Montgomery DR (2007) DIRT: the erosion of civilization. University of California Press, Berkeley, p 285

Naegele, J (1993) See Preface—Zimdahl, RL (1993) Weed Science—a plea for thought

Thill DC, Lish JM, Callihan RH, Bechinski EJ (1991) Integrated weed management—a component of integrated pest management: A critical review. Weed Technol 5:648–656

Thompson PB (1989) Values and food production. J Agric Ethics 2:209–223

Wyse DL (1994) New technologies and approaches the management and sustainable agriculture systems. Weed Technol 8:403–407

Zimdahl RL (1994) Who are you and where are you going? Weed Technol 8:388–391

Index

R. L. Zimdahl, *Weed Science: A Plea for Thought*—Revisited,
SpringerBriefs in Agriculture, DOI: 10.1007/978-94-007-2088-6,
© Robert L. Zimdahl 2012